Cryptography and Cryptanalysis in Java

Creating and Programming Advanced Algorithms with Java SE 17 LTS and Jakarta EE 10

Stefania Loredana Nita
Marius Iulian Mihailescu

Apress®

Cryptography and Cryptanalysis in Java: Creating and Programming Advanced Algorithms with Java SE 17 LTS and Jakarta EE 10

Stefania Loredana Nita
Bucharest, Romania

Marius Iulian Mihailescu
Bucharest, Romania

ISBN-13 (pbk): 978-1-4842-8104-8
https://doi.org/10.1007/978-1-4842-8105-5

ISBN-13 (electronic): 978-1-4842-8105-5

Managing Director, Apress Media LLC: Welmoed Spahr
Acquisitions Editor: Steve Anglin
Development Editor: Laura Berendson
Copy Editor: Kimberly Wimpsett
Editorial Operations Manager: Mark Powers

Cover designed by eStudioCalamar

Cover image by Anna Zakharova on Unsplash (www.unsplash.com)

Distributed to the book trade worldwide by Apress Media, LLC, 1 New York Plaza, New York, NY 10004, U.S.A. Phone 1-800-SPRINGER, fax (201) 348-4505, e-mail orders-ny@springer-sbm.com, or visit www.springeronline.com. Apress Media, LLC is a California LLC and the sole member (owner) is Springer Science + Business Media Finance Inc (SSBM Finance Inc). SSBM Finance Inc is a **Delaware** corporation.

For information on translations, please email booktranslations@springernature.com; for reprint, paperback, or audio rights, please e-mail bookpermissions@springernature.com.

Apress titles may be purchased in bulk for academic, corporate, or promotional use. eBook versions and licenses are also available for most titles. For more information, reference our Print and eBook Bulk Sales web page at www.apress.com/bulk-sales.

Any source code or other supplementary material referenced by the author in this book is available to readers on GitHub (github.com/apress). For more detailed information, please visit www.apress.com/source-code.

Printed on acid-free paper

To our families and to the wonderful readers. May this book be a true inspiration for everyone...and remember: "If you think technology can solve your security problems, then you don't understand the problems, and you don't understand technology."

—*Bruce Schneier*

Table of Contents

About the Authors

Stefania Loredana Nita, PhD, is a lecturer at "Ferdinand I" Military Technical Academy of Bucharest and a software developer and researcher at the Institute for Computers. Her PhD thesis was on advanced cryptographic schemes using searchable encryption and homomorphic encryption. At the Military Technical Academy, she teaches the Formal Languages and Translators and Database Application Development courses. She worked for more than two years as an assistant lecturer at the University of Bucharest where she taught courses on subjects such as advanced programming techniques, simulation methods, and operating systems. Her research activity is in the cryptography field, with a focus on searchable encryption and homomorphic encryption. She is also interested in blockchain, quantum cryptography, machine learning, and artificial intelligence. At the Institute for Computers, she is working on research and development projects that involve cloud computing security, the Internet of Things, and big data. She has authored or co-authored more than 28 papers at conferences and for journals and has co-authored five books. Also, she holds an MSc in software engineering and two BSc degrees in computer science and mathematics.

Marius Iulian Mihailescu, PhD, has worked in the academic and industry sector for more than 15 years. Currently, he is an associate professor (senior lecturer) of engineering and computer science at "Spiru Haret" University, Romania, and as a side job he is a project manager at the Institute for Computers where he is managing different projects using different technologies, such as DevOps, Scrum, Agile, C#, Microsoft SQL Server, Syncfusion, ASP.NET, and VUE. At the university, he has taught several key computer science courses about information security, functional programming, Interne of Things, blockchain, software development methods (Microsoft Azure, Entity Framework, NHibernate, LINQ-to-SQL, UX with DevExpress controls, etc.), and development web applications (HTML 5, CSS 3, Bootstrap, JavaScript, AJAX, NodeJS, VUE, Laravel, mRabbit, ASP.NET, PHP). He has authored or co-authored more than

30 articles in conference proceedings, 25 articles in journals, and six books. For three years he worked as an IT officer at Royal Caribbean Cruises Ltd. where he dealt with IT infrastructures, data security, and satellite communications systems. He received his PhD in 2014, and his thesis was on applied cryptography over biometrics data. He holds two MSc degrees in information security and software engineering, from "Ferdinand I" Military Technical Academy and the University of Bucharest, Romania.

About the Technical Reviewer

Doug Holland is a software engineer and architect at Microsoft Corporation. He holds a master's degree in software engineering from the University of Oxford. Before joining Microsoft, he was honored with the Microsoft MVP and Intel Black Belt Developer awards.

CHAPTER 1

Introduction

In the last decade, technology has rapidly evolved. Statistics show that 64.2 zettabytes of data were generated in 2020 (1 zettabyte is equivalent to 10^{21} bytes or 10^{12} gigabytes), and it is predicted that by 2025, the digital data generated will reach 181 zettabytes [1, 2]. Electronic communication has become an essential part of our lives, and due to its rapid evolution, all manner of security issues have arisen. Because digital messages, in all of their forms, are sent daily over public networks across the world, the need for secure channels and security mechanisms has also increased. Digital devices and communications should have digital signatures that make them easy to authenticate. Modern cryptography provides solutions for all these requirements.

The era in which we are living is considered the "zettabytes era," in which technology allows humans and electronic devices to generate and send information instantly, at any time and any place. Advanced technologies, such as the Internet of Things, fog computing, edge computing, smart vehicles, drones, smart houses, and many other complex software (desktop/web/mobile) solutions or architectures, are evolving so quickly that it is difficult to keep up with security requirements. For example, at the time of this book's writing, there are 160,974 records of vulnerabilities registered on the CVE platform [3]. However, lessons can be learned even from failures, so by analyzing such vulnerabilities, security solutions can be improved.

One of the most important aspects considered when complex systems are designed and implemented is knowledge. In antiquity, the Latins said *Scientia potentia est,* meaning "Knowledge is power" [4]. In the 21s century, this is even more true; information falling into the wrong hands can lead to huge business losses and catastrophic outcomes. Cryptography and information security provide security

© Stefania Loredana Nita and Marius Iulian Mihailescu 2022
S. L. Nita and M. I. Mihailescu, *Cryptography and Cryptanalysis in Java,*
https://doi.org/10.1007/978-1-4842-8105-5_1

mechanisms that can protect information shared between senders and recipients over insecure channels, so that unauthorized users cannot access or alter the transmitted information. Over time, there were encryption systems that were broken by attackers by exploiting vulnerabilities of the systems.

The word *cryptography* comes from the Greek words *kryptos* and *graphein*, meaning "hidden" and "writing," respectively. As its name suggests, the purpose of cryptography is to hide messages from unauthorized individuals and to keep their integrity. Although the study of cryptography has been around only about 100 years, it was used in different forms from ancient times. However, over time there have been various primary methods of hiding secret messages, starting with hieroglyphs, continuing with Caesar's famous cipher, followed by the Vigenère cipher, Hebern's rotor machine, and the famous Enigma machine. Nevertheless, hiding messages was not the only occupation close to cryptography or, rather, information security. Another example is authentication or identity verification; this was often done through seals.

Cryptography is considered an art, especially in its primary phases. The history of cryptography began in ancient Egypt, alongside the art of writing, during a time when humans started organizing in different social groups. This organization led to a natural need of transmitting information only to certain individuals, with the same group, tribe, etc. Early forms of cryptography were hieroglyphs, which started to be used about 4,000 ago by Egyptians—only they recognized the symbols and their meaning. An inscription carved circa 1900 BC contains the first known evidence of cryptography (in some kind). It is located in Egypt nobleman Khnumhotep II's tomb, in the main chamber [5]. In this inscription, some symbols have a different form than usual, and the scribe's intent was not necessary to hide a message; rather, he wanted the symbols to look nobler than usual according to the social status of the deceased. Although the inscription does not hide a message, it contains an altered/transformed form of the original symbols, being the oldest proof of such an approach. Then, cryptography in the ancient world moved to a substitution approach, in which every symbol of an alphabet was replaced by another symbol based on a secret rule. This was happening around 500–600 BC. The next notable cipher was Caesar's cipher. Caesar was a Roman emperor who was communicating with his army generals with encoded messages, using a substitution within the Roman alphabet. Each letter was shifted a certain number of positions in the alphabet, usually three. For example, the correspondent of A was D, of B was E, and so on. This is an important historical cipher that is mentioned often in cryptography literature. The next important achievement in cryptography was in the Middle Ages by Leon Battista Alberti, who implemented polyalphabetic substitution. Two rotating copper disks were used that had the alphabet inscribed on them. Different variations of polyalphabetic substitution ciphers were used, but the most known such cipher is Vigenère. Then in the 19th century, the encryption methods evolved and became more technical. The beginning of modern cryptography starts mainly with the Enigma machine, although the rotors were used a few years before Enigma's invention. Considered unbreakable, the Enigma machine was invented by German engineer Arthur Scherbius at the end of World War I, but it was extensively used in World War II by the German army. Enigma is based on more rotors that work electromechanically, and it scrambles the letters of the alphabet. During World War II, cryptography alongside cryptanalysis evolved quickly and became mathematized. Then modern cryptography continued with symmetric encryption (using a private key for both encryption and decryption) and was followed by asymmetric encryption (where a public key is used for encryption and a private key for decryption) introduced by Diffie and Hellmann in 1976. Since then, different types of encryption systems evolved from each type of cryptography (symmetric or asymmetric).

For a more detailed history or interesting facts about cryptography, you can consult [6] and [7].

The book aims to present the main topics of cryptography, information security, and cryptanalysis from a practical perspective, by providing examples of implementations in Java. The book addresses a large audience, such as security experts, military experts and researchers, ethical hackers, teachers in academia, researchers, software developers, and software engineers, and it can represent a good starting point in developing secure applications, together with [8], [9], [10].

Cryptography and Cryptanalysis

When working with information security and data protection, the concepts of cryptology, cryptography, and cryptanalysis should be clear. These are defined here, as presented in [8, 9]:

- *Cryptology* is defined as the science or art of secret writings; the main goal is to protect and defend the secrecy and confidentiality of the information with the help of cryptographic algorithms.

- *Cryptography* represents the defensive side of cryptology; the main objective is to create and design cryptographic systems and their rules. When we are dealing with cryptography, we can observe a special kind of art, an art that is based on protecting the information by transforming it into an unreadable format, called *ciphertext*.

- *Cryptanalysis* is the offensive side of cryptology; its main objective is to study the cryptographic systems with the scope to provide the necessary characteristics in such a way to fulfill the function for which they have been designed. Cryptanalysis can analyze the cryptographic systems of third parties through the cryptograms realized with them, breaking them to obtain useful information for their business purpose. Cryptanalysts, code breakers, or ethical hackers are the people who in the field of cryptanalysis.

- *Cryptographic primitive* represents a well-established or low-level cryptographic algorithm is used to build cryptographic protocols. Examples of such routines include hash functions or encryption functions.

Book Structure

This book contains 15 chapters, in which the main aspects of classical and modern cryptography are presented. Generally, the chapters will cover the foundation of the presented concept/mechanism/technique from a mathematical perspective and then a practical implementation or use cases in Java. The following chapters are detailed here:

- *Chapter 2, "JDK 17: New Features"*: This chapter will cover the new features of Java 17 and will show some practical examples.

- *Chapter 3, "Roadmap and Vision for Jakarta EE10"*: This chapter will present the basic usage of Jakarta EE and explain how security mechanisms can be integrated.

- *Chapter 4, "Java Cryptography Architecture"*: This chapter presents the built-in functions of Java that can be used in cryptography. These are encapsulated in Java's cryptography application programming interface (API) called Java Cryptography Architecture (JCA).

- *Chapter 5, "Classical Cryptography"*: This chapter will describe classic enciphering techniques. These ciphers use basic mathematical functions but represent a good starting point in understanding the purpose of cryptography and its basic rules. For each classic algorithm, the mathematical description will be presented followed by the implementation in Java.

- *Chapter 6, "Formal Techniques for Cryptography"*: This chapter is focused on the formal aspects of cryptography. It will present, without going into too many technical details, the main mathematical elements used in cryptography that are mandatory to understand; it will also define some specific terms of cryptography.

- *Chapter 7, "Pseudorandom Generators"*: Randomness is one of the most important concepts used in cryptography. This chapter will present what pseudorandom generators are and why are they important in cryptography. One section of the chapter is dedicated to the `Java.util.Random` class.

- *Chapter 8, "Hash Functions"*: This chapter will explore how hash values can be generated for different types of data and will present and implement some important hash functions, like the MD or SHA families.

- *Chapter 9, "Symmetric Encryption Algorithms"*: This chapter will explain what symmetric encryption is and will present two of the most important and used symmetric cryptosystems: AES and DES.

- *Chapter 10, "Asymmetric Encryption Schemes"*: This chapter will highlight the differences between symmetric and asymmetric cryptography and describe and implement two of the most important cryptosystems of this category: RSA and ElGamal. Also, the chapter presents the Merkle-Hellman Knapsack System, which is interesting due to its approach.

- *Chapter 11, "Signature Schemes"*: This chapter will explain what cryptographic signature schemes are, why are they so important, and where can they be used. Then, it will present some important signature schemes, such as ElGamal.

- *Chapter 12, "Identification Schemes"*: This chapter will explain what other important cryptographic schemes, namely, identification schemes, are; why are they so important; and where can they be used. Then, it will present some important identification schemes, such as CVE.

- *Chapter 13, "Lattice-Based Cryptography and NTRU"*: This chapter will present the main concepts of lattice-based cryptography and will describe the NTRU encryption system that is included in this branch of cryptography.

- *Chapter 14, "Advanced Encryption Schemes"*: This chapter will cover two relatively new encryption techniques, namely, searchable and homomorphic encryption. The second one is very exciting, especially the fully homomorphic encryption, which is considered the holy grail of cryptography.

- *Chapter 15, "Cryptography tools"*: This chapter will discuss cryptography tools that can be used to check the correctness of their implementations, such as CrypTool or OpenSSL.

Conclusion

This first chapter covered the objectives of the book and highlighted the need for cryptography and information security. Then it explained what cryptology, cryptography, and cryptanalysis mean. Finally, the chapter described each upcoming chapter so you can find information of interest quickly.

References

[1].   Volume of data/information created, captured, copied, and consumed worldwide from 2010 to 2025, https://www.statista.com/statistics/871513/worldwide-data-created/

[2].   Zettabyte era, https://en.wikipedia.org/wiki/Zettabyte_Era

[3].   CVE, https://cve.mitre.org/

[4]. Scientia potentia est, https://en.wikipedia.org/wiki/
Scientia_potentia_est

[5]. Sidhpurwala, H. (2013). "A Brief History of Cryptography".
Available online: https://www.redhat.com/en/blog/brief-
history-cryptography

[6]. History of cryptology. Available online: https://www.britannica.
com/topic/cryptology/History-of-cryptology

[7]. Damico, T. M. (2009). A brief history of cryptography. Inquiries
Journal, 1(11).

[8]. Mihailescu, M. I., & Nita, S. L. (2021). Pro Cryptography and
Cryptanalysis: Creating Advanced Algorithms with C# and
.NET. Apress.

[9]. Mihailescu, M. I., & Nita, S. L. (2021). Pro Cryptography and
Cryptanalysis with C++ 20: Creating and Programming Advanced
Algorithms. Apress.

[10]. Mihailescu, M. I., & Nita, S. L. (2021). Cryptography and
Cryptanalysis in MATLAB. Apress.

CHAPTER 2

JDK 17: New Features

The Java Platform—developed by Oracle—includes two components: Java Standard Edition (Java SE) and the Java Development Kit (JDK). Java SE is a computing platform that is used to create software applications by developing and deploying portable code for environments such as desktops and servers. Java SE includes different application programming interfaces (APIs) and the Java Class Library. Examples of packages included in Java SE are `java.io`, `java.math`, `java.util`, etc., and it includes APIs such as Applet, AWT, Collections, Swing, JDBC, etc. You can find more about Java SE and its editions at [1].

JDK is the Java development environment, which has two components: Java Language Specification (JLS) and Java Virtual Machine Specification (JVMS). At the moment of writing this book, the latest version of JDK is 17.0.1, released on October 19, 2021, but Java 17 and the Long Term Support (LTS) were released on September 14, 2021. Being an LTS version, JDK 17 will receive support and updates until at least September 2024 [2]. Some great resources for Java developers are at [3], which will provide soon a built-in Java compiler to test the examples, and [4], which is the official Oracle documentation.

In JDK 17, several changes include new features, feature removals, and feature deprecations. You can find the complete list of changes in Java 17 at [5], [6], and [7]. The changes are encoded as "JEP" followed by a number, where JEP means JDK Enhancement Proposals. Some of the main changes are listed in the following lists.

New features:

- JEP 306: Restore Always-Strict Floating-Point Semantics

- JEP 356: Enhanced Pseudorandom Number Generators

- JEP 382: New macOS Rendering Pipeline

- JEP 391: macOS/AArch64 Port

© Stefania Loredana Nita and Marius Iulian Mihailescu 2022
S. L. Nita and M. I. Mihailescu, *Cryptography and Cryptanalysis in Java*,
https://doi.org/10.1007/978-1-4842-8105-5_2

- JEP 403: Strongly Encapsulate JDK Internals

- JEP 406: Pattern Matching for `switch` (Preview)

- JEP 409: Sealed Classes

- JEP 412: Foreign Function & Memory API (Incubator)

- JEP 414: Vector API (Second Incubator)

- JEP 415: Context-Specific Deserialization Filters

Removals:

- JEP 407: Remove RMI Activation

- JEP 410: Remove the Experimental AOT and JIT Compiler

Deprecations:

- JEP 398: Deprecate the Applet API for Removal

- JEP 411: Deprecate the Security Manager for Removal

Note that in the following examples, the Eclipse environment should be configured for use with Java 17. We'll now elaborate on the JEPs of interest to the field of cryptography.

JEP306, "Restore Always-Strict Floating-Point Semantics." This new feature brings back the semantics before introducing the strict and default floating-point modes in Java SE 1.2. The purpose of this modification is to make it easy to work with libraries for numeric usages, such as `java.lang.Math` or `java.lang.StrictMath`. The variations of the types of calculations (strict, strict floating-point) were due to hardware constraints in that some processors suffered from overheating issues while performing strict computations. However, nowadays processors have overcome such constraints; therefore, strict computations can be used safely without restrictions. You can learn more about JEP306 at [8].

JEP356, "Enhanced Pseudorandom Number Generators." This one is of interest for cryptography use. This new feature includes new interface types and new implementations for pseudorandom number generators (PRNGs), which are included in the `java.util.random` package. The new implementations contain jumpable PRNGs and a completely new class for splittable PRNGs, called LXM. The purpose of this feature is to make it easier to use PRNGs and to eliminate two weaknesses in the `SplittableRandom` class discovered in 2016.

In Listing 2-1 are printed available PRNGs with some information about them: their type (arbitrary jumpable, hardware, jumpable, leapable, splittable, statistical, stochastic, or streamable) and the number of bits used to maintain the state of the seed. After that, lines 28 and 29 show how PRNGs are instantiated in the new Java 17, while line 31 shows the old instantiation of the Random class. Lines 33, 34, and 36 show the default representation for the new implementations and the shortcut for the default representation, respectively. Figure 2-1 shows the results.

Listing 2-1. Using PRNGs in Java 17

```
1    import java.util.Random;
2    import java.util.random.RandomGenerator;
3    import java.util.random.RandomGeneratorFactory;
4    import java.util.stream.Stream;
5
6    public class JEP356Example {
7
8        public static void main(String[] args) {
9
10           Stream<RandomGeneratorFactory<RandomGenerator>> allPRNGs =
              RandomGeneratorFactory.all();
11           allPRNGs.map(prng -> prng.name() + " [ Group: "  + prng.
              group() + "; "
12                       + (prng.isArbitrarilyJumpable() ? " arbitrary-
                         jump" : "")
13                       + (prng.isHardware()? " hardware" : "")
14                       + (prng.isJumpable() ? " jump" : "")
15                       + (prng.isLeapable()? " leap" : "")
16                       + (prng.isSplittable() ? " split" : "")
17                       + (prng.isStatistical()? " statistical" : "")
18                       + (prng.isStochastic()? " stochastic" : "")
19                       + (prng.isStreamable() ? " stream" : "")
20                       + "; noOfBits: "+ prng.stateBits()
21                       + "]"
22                       ).sorted().forEach(System.out::println);
23
```

```
24                  System.out.println("\n*****\n");
25
26                  RandomGenerator prng1 = RandomGeneratorFactory.
                    of("Random").create(45);
27                  System.out.println("prng1 - " + prng1.getClass());
28                  RandomGenerator prng2 = new Random(45);
29                  System.out.println("prng2 - " + prng2.getClass());
30                  RandomGenerator prng3 = RandomGeneratorFactory.
                    getDefault().create(45);
31                  System.out.println("prng3 - " + prng3.getClass());
32                  RandomGenerator prng4 = RandomGenerator.getDefault();
33                  System.out.println("prng4 - " + prng4.getClass());
34          }
35     }
```

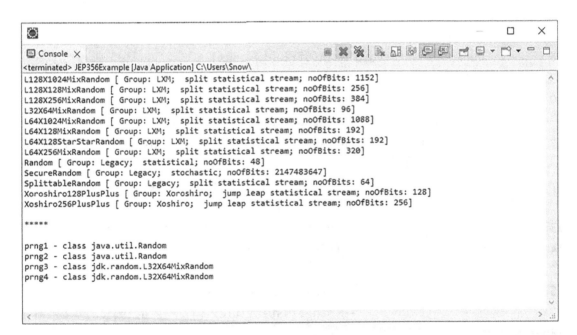

Figure 2-1. *The result of Listing 2-1*

In Figure 2-1, the `Legacy` category is for the old implementation of PRNGs, while `LXM` and `Xoroshiro` are newly introduced. Note that the new implementations are not thread-safe; namely, the same instance can be shared between more threads (although this is not a common practice while developing applications). Instead, `java.util.Random` and `java.security.SecureRandom` are thread-safe. You can learn more about JEP 356 at [9].

JEP403, "Strongly Encapsulate JDK Internals." This is one of the features that enhances the security of the JDK, as the internal APIs cannot be accessed with the `--illegal-access` option at the command line anymore; this is because in JDK 17 all internal elements are now strongly encapsulated. The exceptions are crucial APIs, such as `sun.misc.Unsafe`. The purpose of this feature is to encourage developers to use standard APIs instead of internal APIs. This change increases the security and the maintainability of the JDK. On the official page of this JEP, the developers give some examples of pieces of code that will not work anymore in JDK 17 and provide explanations for each example. You can learn more about JEP 403 at [10].

JEP406, "Pattern Matching for switch (Preview)." By far, using pattern matching in `switch` expressions and statements will be easier and more elegant. In addition, this will save a lot of lines of code. Listing 2-2 is an example of pattern matching for `switch` that can be used in cryptography. Consider three classes that implement the RSA, AES, and ElGamal algorithms (lines 1–11); the user should introduce only the secret key or the pair of the secret key and public key corresponding to the chosen encryption system's type. Note that for the following example, the usage of preview language features should be enabled in order to work properly.

Listing 2-2. Example of Pattern Matching for switch

```
1    class RSA {
2
3    }
4
5    class AES {
6
7    }
8
9    class ElGamal {
10
11    }
```

```
12
13   public class JEP406Example {
14
15       static void TypeOfKeys(Object o) {
16           if (o instanceof RSA ) {
17               System.out.println("Two keys needed. Type the secret on
                     the first line. Type the public key on the second line.");
18           } else if (o instanceof AES) {
19               System.out.println("One key needed. Type the secret
                     key on the first line.");
20           } else if (o instanceof ElGamal) {
21               System.out.println("Two keys needed. Type the secret on
                     the first line. Type the public key on the second line.");
22           }
23       }
24
25       static void TypeOfKeysPatternSwitch(Object o) {
26           if(o == null)
27               throw new NullPointerException();
28           else
29           {
30               switch (o.getClass().toString()) {
31                       case null -> throw new NullPointerException();
32                       case "RSA2" -> System.out.println("Two keys
                             needed. Type the secret on the first line. Type
                             the public key on the second line.");
33                       case "AES2" -> System.out.println("One key
                             needed. Type the secret key on the first line.");
34                       case "ElGamal2" -> System.out.println("Two keys
                             needed. Type the secret on the first line. Type
                             the public key on the second line.");
35                       default -> System.out.println("Pick an
                             encryption system");
36               }
37           }
38       }

14
```

```
39
40        public static void main(String[] args) {
41              TypeOfKeys(new RSA());
42              TypeOfKeysPatternSwitch(new RSA());
43
44              System.out.println("\n***\n");
45
46              TypeOfKeys(new AES());
47              TypeOfKeysPatternSwitch(new AES());
48        }
49    }
```

In Listing 2-2, two approaches were used to print the message for the user: the regular one, in the function TypeOfKeys (lines 16–27), and the version with pattern matching for the switch statement (lines 29–45). Note that the result should be the same for the same type of object that is passed to the two functions. You can check the result in Figure 2-2, and you can learn more about JEP 406 at [11].

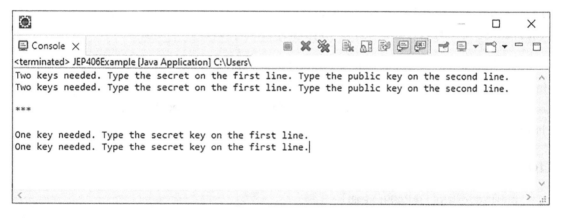

Figure 2-2. *The result of using pattern matching for switch*

JEP409, "Sealed Classes." These types of classes are important in object-oriented programming because they can restrict the list of classes or interfaces that are extensions or implementations of them, giving more control to the code that implements them. For example, consider there is a fixed collection of encryption systems stored within an enum. As there is a fixed number of encryption systems, a default clause wouldn't be necessary when switching between them; see Listing 2-3.

Listing 2-3. Sketch for Cryptosystems Stored as enum

```
enum Cryptosystem { AES, RSA, ElGamal }

Cryptosystem cipher = ...
switch (cipher) {
  case AES: ...
  case RSA: ...
  case ElGamal: ...
}
```

In situations in which a given number of kinds of values should be managed (instead of managing a set of values), a hierarchy of classes can be used for listing these kinds of values. For the previous enum example, this would be transformed into Listing 2-4.

Listing 2-4. Converting the enum Example to a Class Hierarchy

```
interface Cryptosystem { ... }
final class AES implements Cryptosystem { ... }
final class RSA implements Cryptosystem { ... }
final class ElGamal implements Cryptosystem { ... }
```

However, superclasses should be widely accessible but not widely extensible. Sealed classes were introduced to give more control over how extensibility and implementation are used. The previous example can be modified by using sealed classes as shown in Listing 2-5.

Listing 2-5. Sealed Classes

```
1    public class JEP409Example {
2
3        static void Test(Cryptosystem c) {
4            if (c instanceof AES_Cryptosystem)
5                System.out.println("AES chosen");
6            else
7                if (c instanceof RSA_Cryptosystem)
8                    System.out.println("RSA chosen");
```

```
 9                      else
10                              throw new RuntimeException("Unknown instance
                                of Cryptosystem.");
11
12           }
13           public static void main(String[] args) {
14                   AES_Cryptosystem aes = new AES_Cryptosystem();
15                   RSA_Cryptosystem rsa = new RSA_Cryptosystem();
16
17                   Test(aes);
18                   Test(rsa);
19           }
20   }
```

In Listing 2-5, three classes are used: Cryptosystem, AES_Cryptosystemn, and RSA_ Cryptosystem. Note that to use them as sealed classes and classes that extend the sealed class, each class must be defined in its own file. To prevent further extensions, the classes that extend the sealed class should be declared as final. Here is the definition for the three classes used in the example:

```
public abstract sealed class Cryptosystem
    permits AES_Cryptosystem, RSA_Cryptosystem {

    String encryptMessage() { return null;}
    String decryptMessage() { return null; }
}

public final class AES_Cryptosystem extends Cryptosystem {

}

public final class RSA_Cryptosystem extends Cryptosystem {

}
```

You can find more examples of sealed classes at [12].

JEP412, "Foreign Function & Memory API (Incubator)." With this new feature, Java code can work with foreign code and data from outside of the Java runtime, other than C. In cryptography, this can be useful, especially when working with mathematical tools.

For example, Matlab code can be used for mathematical functions or algorithms that are already implemented, such as statistical functions. Using the foreign code is safe and efficient. You can learn more about JEP412 at [13].

JEP414, "Vector API (Second Incubator)." Using this feature achieves better performance and implementation for vector instructions compared to equivalent scalar computations. The improved API provides features such as translating byte vectors to Boolean arrays, or vice versa. You can learn more about JEP412 at [14].

JEP415, "Context-Specific Deserialization Filters." This is another example of a feature that is great for cryptographic purposes. The motivation behind this feature lies in the fact that data that is received from different sources is untrusted because in the world of the Internet, often the source is unknown, untrusted, or unauthenticated. Therefore, when untrusted data is deserialized, it can be dangerous. For example, an attacker can exploit the vulnerabilities behind serialized/deserialized data injecting malicious behavior. With this feature, a filter factory for Java Virtual Machine (JVM) was introduced that can be configured. When it is instantiated as an object of the `ObjectInputStream` type, the JVM filter factory is invoked, and the result is used to initialize each filter per stream. You can learn more about JEP415 at [15].

Conclusion

This chapter presented the main features added in Java 17. By far one of the most exciting added features is the new pseudorandom generator implementations and pattern matching for `switch` instructions. The chapter presented some use cases for cryptography.

References

[1]. Java SE at a Glance, `https://www.oracle.com/ro/java/technologies/java-se-glance.html`

[2]. Java Downloads, `https://www.oracle.com/java/technologies/downloads/#java17`

[3]. `https://dev.java/`

[4].   JDK 17 Documentation, https://docs.oracle.com/en/java/
         javase/17/

[5].   JDK 17 Release Notes, https://www.oracle.com/java/
         technologies/javase/17-relnote-issues.html

[6].   Java® Platform, Standard Edition & Java Development Kit Version
         17 API Specification, https://docs.oracle.com/en/java/
         javase/17/docs/api/index.html

[7].   JDK 17.0.1 General-Availability Release, https://jdk.
         java.net/17/

[8].   JEP 306: Restore Always-Strict Floating-Point Semantics, https://
         openjdk.java.net/jeps/306

[9].   JEP 356: Enhanced Pseudorandom Number Generators, https://
         openjdk.java.net/jeps/356

[10].   JEP 403: Strongly Encapsulate JDK Internals, https://openjdk.
         java.net/jeps/403

[11].   JEP 406: Pattern Matching for switch (Preview), https://openjdk.
         java.net/jeps/406

[12].   JEP 409: Sealed Classes, https://openjdk.java.net/jeps/409

[13].   JEP 412: Foreign Function & Memory API (Incubator), https://
         openjdk.java.net/jeps/412

[14].   JEP 414: Vector API (Second Incubator), https://openjdk.java.
         net/jeps/414

[15].   JEP 415: Context-Specific Deserialization Filters, https://
         openjdk.java.net/jeps/415

CHAPTER 3

Roadmap and Vision for Jakarta EE 10

Jakarta EE 10 will represent the first major release of Jakarta EE, and its goal is to deliver a specific set of specifications for Jakarta EE technologies.

By late 2021, the key themes that have been proposed by the community for shaping Jakarta EE 10 are as follows:

- Aligning with Java Standard Edition (SE)

- Filling the gaps in the standardization

- Aligning with contexts and dependency injection (CDI)

Starting with Java EE 8, Jakarta Security is a new API with three authentication procedures: Basic, Form, and an extension of Form that perfectly suits the process of implementation within Java Server Faces (JSF).

With Jakarta EE 10, the goal is to add new authentication mechanisms. The first mechanism on the list is represented by Client-Cert and Digest. In this way, Jakarta Security will provide a total replacement of the authentication mechanisms that are supported by Java Servlet. Another goal is to offer support for OpenID, OAuth, and JSON Web Token (JWT). JWT is an interesting topic that is currently under discussion due to the process of adding it to the MicroProfile project.

Jakarta Security supports different annotations, such as @RolesAllowed and @RunAs, with support for further extensions. At the moment, @RolesAllowed is supported only by EJB in Jakarta EE. The annotation will throw an exception if access is denied by a bean method.

Related to the authentication mechanisms, @RolesAllows will trigger an authentication mechanism when access is denied by a Jakarta REST resource method.

© Stefania Loredana Nita and Marius Iulian Mihailescu 2022
S. L. Nita and M. I. Mihailescu, *Cryptography and Cryptanalysis in Java*,
https://doi.org/10.1007/978-1-4842-8105-5_3

The Jakarta EE 10 platform will include a serious set of components; Table 3-1 shows their status.

Table 3-1. *Jakarta EE 10 Web Profile*

Component	Status
Authentication 3.0	Updated
Concurrency 3.0	Updated
CDI 4.0	Updated
Expressions Language 5.0	Updated
Faces 4.0	Updated
Security 3.0	Updated
Servlet 6.0	Updated
Standard Tag Libraries 3.0	Updated
Persistence 3.1	Updated
Server Pages 3.1	Updated
WebSocket 2.1	Updated
Bean Validation 3.0	Not updated
Debugging Support 2.0	Not updated
Enterprise Beans Lite 4.0	Not updated
Managed Beans 2.0	Not updated
Transactions 2.0	Not updated
RESTful Web Services 3.1	Updated
JSON Processing 2.1	Updated
JSON Binding 2.1	Updated
Annotations 2.0	Not updated
Interceptors 2.0	Not updated
Dependency Injection 2.0	Not updated
CDI Lite 4.0	Updated/new
Config 1.0	New

User-friendly authentication modules are one of the most important features handled by Jakarta Security within Jakarta EE 10; they enable authorization rules that can be easily implemented.

These authorization modules are designed to work as low-level, portable mechanisms. The steps used to develop and install these modules have become more stable, making them more useful for business applications.

The specifications that are available with the Jakarta EE 10 platform and Web Profile for business applications are as follows, according to the Jakarta EE 10 Release Plan (`https://eclipse-ee4j.github.io/jakartaee-platform/jakartaee10/JakartaEE10ReleasePlan`):

- Jakarta EE Platform 10 (updated feature)

- Jakarta EE Web Profile 10 (updated feature)

- Jakarta Activation 2.1 (updated feature)

- Jakarta Annotations 2.1 (updated feature for Web Profile)

- Jakarta Authentication 3.0 (updated feature for Web Profile)

- Jakarta Authorization 2.1 (updated feature)

- Jakarta Batch 2.1 (updated feature)

- Jakarta Bean Validation 3.0 (feature for Web Profile)

- Jakarta Concurrency 3.0 (updated feature for Web Profile)

- Jakarta Connectors 2.1 (updated feature)

- Jakarta Contexts and Dependency Injection 4.0 (updated feature)

- Jakarta Debugging Support for Other Languages 2.0 (feature for Web Profile)

- Jakarta Dependency Injection 2.0 (feature for Web Profile)

- Jakarta Enterprise Beans 4.0

- Jakarta Enterprise Beans 4.0 Lite (feature Web Profile)

- Jakarta Enterprise Web Services 2.0 (optional feature)

- Jakarta Expression Language 5.0 (updated feature for Web Profile)

- Jakarta Interceptors 2.0 (feature for Web Profile)

- Jakarta JSON Binding 3.0 (updated feature for Web Profile)

- Jakarta JSON Processing 2.1 (updated feature Web Profile)

- Jakarta Mail 2.1 (updated feature)

- Jakarta Managed Beans 2.0 (Web Profile)

- Jakarta Messaging 3.1 (updated feature)

- Jakarta Persistence 3.1 (updated feature for Web Profile)

- Jakarta RESTful Web Services 3.1 (updated feature for Web Profile)

- Jakarta Security 3.0 (updated feature for Web Profile)

- Jakarta Server Faces 4.0 (updated feature for Web Profile)

- Jakarta Server Pages 3.1 (updated feature for Web Profile)

- Jakarta Servlet 6.0 (updated feature for Web Profile)

- Jakarta SOAP with Attachments 3.0 (optional feature)

- Jakarta Standard Tag Library 3.0 (feature for Web Profile)

- Jakarta Transactions 2.0 (feature for Web Profile)

- Jakarta WebSocket 2.1 (feature for Web Profile)

- Jakarta XML Binding 4.0 (optional feature)

- Jakarta XML Web Services 4.0 (optional feature)

At the time of this book's writing, the tool that contains portable modules for low-level authorization had not yet been incorporated into Jakarta Security 1.0, as it had not yet been fully evaluated. For example, to have a solution that identifies the scope of an application, checking for the bridging role against an external service is the best strategy, instead of taking all the roles and assigning them when the invocation call is authenticated. Such an example is shown here:

```
1    import java.io.UnsupportedEncodingException;
2    import java.security.KeyPair;
3    import java.security.KeyPairGenerator;
4    import java.security.NoSuchAlgorithmException;
```

```
5    import java.security.NoSuchProviderException;
6    import java.security.PrivateKey;
7    import java.security.PublicKey;
8    import java.security.SecureRandom;
9
10   @ApplicationScoped
11   public class ApressAuthorizationModule {
12
13       @Inject
14       SecurityConstraints apressSecurityConstraints
15
16       @Inject
17       ApressService apressService;
18
19       @PostAuthenticate
20       @PreAuthorize
21       @ByRole
22       public Boolean appressLogic(
23           Caller apressCaller, Permission apressReqPermission) {
24
25           return apressSecurityConstraints.getRequiredRoles(apressReq
             Permission)
26                   .stream()
27                   .anyMatch(apressAccessRole -> apressService.
                     isInRole(apressCaller, apressAccessRole));
28       }
29
30   }
```

Jakarta CDI represents the main component model within the platform, and many components are being aligned with it already. The important components include the following:

- The @Transactional annotation gets a new definition in Jakarta Transactions.

- The managed bean annotations is being deprecated in JSF.

- CDI has been developed and improved in Jakarta Security.

The following components related to Jakarta EE 10 take full advantage of CDI:

- When CDI is enabled, the features of the Jakarta Enterprise Beans programming model will be more flexible. Because of the modernization process within Jakarta EE 10, the features are converted to specifications where they fit perfectly. Here are some examples:

    - Definitions for @Lock, @Schedule, @Asynchronous, and @MaxConcurrency are included. Security mechanisms have also been improved, and special mechanisms are dedicated to working with these annotations.

    - Another important definition that is included is related to the message-driven bean-style @MessageListener within Jakarta Messaging.

    - Jakarta Security will have two annotations that will need to deal with it: @RolesAllowed and @RunAs.

- Most of the additional mechanisms are developed in such a way that they can be injected with the help of CDI to provide deprecations of their own features with respect for CDI and easy integration within CDI. These mechanisms can be summarized as follows:

    - Jakarta RESTful Web Services

    - Jakarta Batch

    - Jakarta Concurrency

Other new potential specifications within Jakarta EE 10 are the following:

- Jakarta NoSQL and Eclipse JNoSQL for giving access to Jakarta EE applications

- Jakarta Model-View-Controller (MVC)

Next, looking at the future of Jakarta EE 10, there are new challenges, but they will not be covered. Most of the components provide experimentation together with analysis and development. Those features are summarized as follows:

- Jakarta Servlet or a variant much lighter

- Jakarta Persistence, with support for reactive NIO Java Database Connectivity (JDBC)

- Jakarta Restful Web Services

- Jakarta MVC

The Java platform and Jakarta EE 10 provide a number of important built-in providers for implementing basic and advanced security services that are used within different types of projects. The main components that work in both Java Platform and Jakarta EE 10 can be summarized as follows:

- *Java Cryptographic Extension (JCE)*: JCE represents an extension from Sun Microsystems for providing encryption and decryption operation for data blocks. JCE represents an important component of the Java Cryptography Architecture (JCA) implementation.

- *Java Secure Socket Extension (JSSE)*: One of the most important data protocols is represented by the Secure Sockets Layer (SSL), which provides data integrity by providing encryption. JSSE has a standard interface and proper implementation for the SSL protocol.

- *Java Authentication and Authorization Service (JAAS)*: JAAS offers an implementation for user authentication. It provides a wide range of login mechanisms within a specific architecture and a dedicated API.

Good references specialized for Jakarta EE are [1-3] and generally for Java are [4-7].

Conclusion

In this chapter, we provided a short overview of the main components and challenges covered by the new Jakarta EE 10 and how these components will impact the developing process of the software applications and the implementation of security mechanisms.

An important aspect at this stage of development is represented by the new security authentication mechanisms that have been released. You should now be able to describe and understand the main challenges and components within Jakarta EE 10.

References

[1]. Juneau, Josh. *Jakarta EE Recipes: A Problem-Solution Approach, Apress*, 2020.

[2]. Manelli, Luciano, and Giulio Zambon. *Beginning Jakarta EE Web Development: Using JSP, JSF, MySQL, and Apache Tomcat for Building Java Web Applications.* Apress, 2020.

[3]. Saeed, Luqman. *Introducing Jakarta EE CDI Contexts and Dependency Injection for Enterprise Java Development.* Apress : Imprint: Apress, 2020.

[4]. Horstmann, Cay S. *Core Java.* Eleventh edition, Pearson, 2019.

[5]. Bloch, Joshua. *Effective Java.* Third edition, Addison-Wesley, 2018.

[6]. McLaughlin, Brett, et al. *Head First Object-Oriented Analysis and Design.* 1st ed, O'Reilly, 2007.

[7]. Sierra, Kathy, and Bert Bates. *Head First Java.* 2nd ed, O'Reilly, 2005.

CHAPTER 4

Java Cryptography Architecture

The Java Cryptography Architecture (JCA) framework offers capabilities for working with cryptography algorithms and offers methods for a variety of cryptography primitives. The cryptography algorithms and methods are implemented and offered through the `java.security` package.

The Java platform is designed in such a way that the most important aspects and components that characterize security are included in an easy and modern approach for developers. The following are the most important components:

- Secure communication

- Language safety

- Public key infrastructure

- Authentication

- Access control

JCA represents only a small part of the entire Java platform, which has a provider and a set of APIs for different cryptography primitives, such as digital signatures, hash functions, validation of certificates, encryption algorithms/primitives, cryptography key generation functions, key management, and algorithms for generating random numbers.

The provider was designed as architecture with respect for the following principles: \

- *Extensibility of the algorithms*: Java is quite flexible when it comes to providing its proper cryptography algorithms or custom providers' and third-party libraries/frameworks. An important aspect that needs to be taken into consideration when developing applications and using third-party frameworks and libraries is to make sure that the

© Stefania Loredana Nita and Marius Iulian Mihailescu 2022
S. L. Nita and M. I. Mihailescu, *Cryptography and Cryptanalysis in Java*,
https://doi.org/10.1007/978-1-4842-8105-5_4

implementation is tested and standardized accordingly. Everything that is outside of the standardization process requires more attention as a developer.

- *Assuring the interoperability and implementing proper support*: As a programming language and developing technology for software applications, Java guarantees interoperability, which is an important criterion to satisfy even the most drastic requirements, especially when you are dealing with different platforms (e.g., Windows ⟷ Ubuntu ⟷ Mac) for the same applications.

- *Guaranteeing the independence of the cryptography algorithms implementation*: When you are dealing with the implementation process for security algorithms, there is no need to implement the algorithms from scratch. Thus, developers and software engineers can use the security libraries and make proper requests for the cryptography algorithms/primitives that they want to implement. The algorithms are already implemented and standardized accordingly.

JCA when used with JDK has two main components:

- The first component introduces the definitions and provides support for different cryptographic services. Different contributors provide different versions of implementations. The framework provides the following important packages:

    - `java.security`

    - `javax.crypto`

    - `javax.crypto.spec`

    - `javax.crypto.interfaces`

- The second component provides the actual implementation of the cryptography algorithms/primitives. The providers are as follows:

    - `Sun`

    - `SunRsaSign`

    - `SunJCE`

Architecture and Design Principles

As we mentioned, assuring independent implementation and interoperability, along with the independence and extensibility of algorithms, plays an important role and represents the main principles taken into consideration when JCA was designed and implemented.

As an example, you can use a digital signature or hash function without considering how the algorithms were implemented. So, you don't have to worry about the details of the implementation or the algorithms that represent the foundation for the concepts.

One of the fundamental questions that comes up is, how is the independence of the algorithm fulfilled? The main thing that a developer has to do is to define the cryptography engines or services along with the definition of the classes that offer capabilities and different functionalities specific to those cryptographic engines. The name of the concepts, according to some references guides [1], is known as *engine classes*. There are a few examples that help in this area, shown here:

- `Cipher` class

- `KeyPairGenerator` class

- `KeyFactory` class

- `MessageDigest` class

- `Signature` class

Another question that will come up is, how is independence implemented? The Cryptographic Service Provider (CSP) represents a package or a set of packages that provides implementations for a variety of cryptographic services, such as digital signatures algorithms, hash algorithms, and key management algorithms. Simply stated, interoperability in information security is the ability of different implementations to achieve their goals by working with each other, exchanging cryptography keys between them, or being able to check each other's cryptography keys.

The final question is, what is the importance and meaning of algorithm extensibility? The answer relies in the simplicity of how the engine classes are added in a very easy and modern way.

The base class `java.security.Provider` comprises all the security providers. For each CSP we will have an instance of this class. The class will offer the following:

- A list with all the security algorithms and security services that are implemented

- The provider name

To use JCA, you simply specify the request for a certain object and its type. With every JDK that you install, one or more providers will be installed and set by default.

Let's see an example of JCA in action. The first thing you need to do is to make a simple request for a certain type of object, such as `MessageDigest`, and to invoke or call a certain algorithm/service, such as the SHA-512 algorithm.

```
message_hash = MessageDigest.getInstance("SHA-512");
message_hash = MessageDigest.getInstance("SHA-384",
"ImplementedProvider1");
```

Note that the previous lines of code are just an example snippet. In real use, `ImplementedProvider1` should be replaced with a valid provider. For example, a provider for SHA-384 is `SUN`. Java enables users to determine the providers available by using the following code:

```
Provider[] providers = Security.getProviders();
```

Figure 4-1 and Figure 4-2 show the requesting process for the SHA-512 algorithm to hash a message. The ordering process shows three different ways of how a hash algorithm is invoked and how a provider can be used as a source of the implementation. Based on the architectures shown in Figure 4-1 and Figure 4-2, the developers can add any other components (e.g., services, interfaces, third-party functions, etc.) and develop new architectures based on them.

Returning to the independence of the algorithms, their definition will achieve the proper goal of providing the implementation of providers that are compliant with the defined interfaces. This is known as an *application programming interface* (API), and it will guarantee that all applications will use it to access a specific type of service.

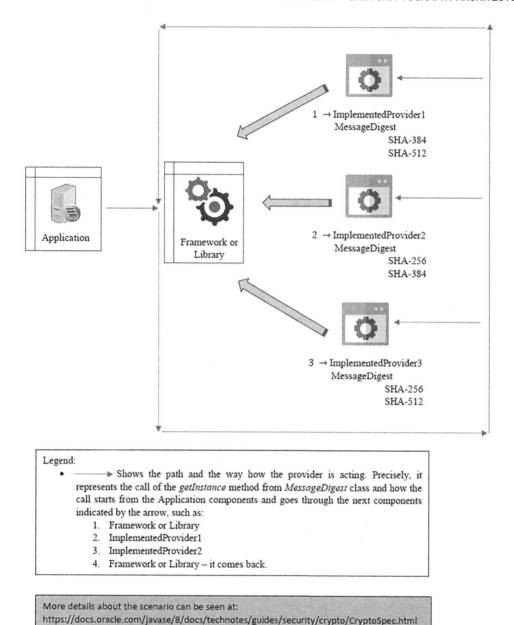

Figure 4-1. *Searching process for provider*

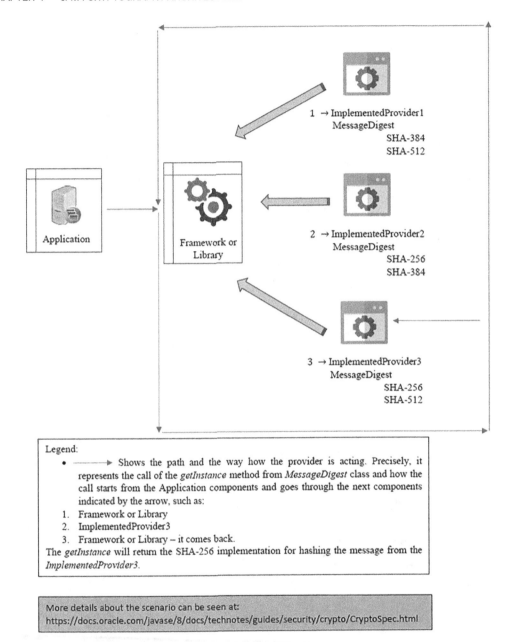

Legend:

- ──────▶ Shows the path and the way how the provider is acting. Precisely, it represents the call of the *getInstance* method from *MessageDigest* class and how the call starts from the Application components and goes through the next components indicated by the arrow, such as:

1. Framework or Library
2. ImplementedProvider3
3. Framework or Library – it comes back.

The *getInstance* will return the SHA-256 implementation for hashing the message from the *ImplementedProvider3*.

More details about the scenario can be seen at:
https://docs.oracle.com/javase/8/docs/technotes/guides/security/crypto/CryptoSpec.html

Figure 4-2. *Requesting a specific provider*

JCA Classes and Algorithms

This section presents the classes and algorithms within JCA.

Algorithms and Engine Classes

The *engine class* is special because it offers an interface for a certain type of cryptography service. The following operations and components are offered by an engine class:

- *Objects*: The objects contain the encrypted data and have the capability to move them to certain and upper abstraction layers. Such examples include certificates or cryptography key containers.

- *Cryptography primitive operations*: These include encryption/ decryption operations, digital signatures, hashing messages, etc.

- *Generators or convertors*: These include mechanisms used for different cryptography concepts/mechanisms, such as cryptography keys or various parameters for cryptography algorithms.

You can find a comprehensive list with examples and case studies at [1]. Table 4-1 lists the most important ones.

Table 4-1. *Engine Classes and Their Purpose*

Engine Class	Purpose
KeyPairGenerator	This generates a pair of public and private keys for a certain algorithm.
KeyGenerator	For an algorithm, the class will generate new secret keys.
KeyAgreement	The instance of the class is used by one or two parties to establish a cryptographic key to be used within a specific cryptographic operation.
Cipher	The class will use the keys (generated for encryption and decryption data). The class provides a set of various algorithms, such as asymmetric encryption, symmetric encryption, and password-based encryption.
MessageDigest	This provides functionalities for computing the hash value for a certain message.
SecureRandom	This provides functionalities for generating random numbers.

Interfaces and Main Classes

Table 4-2 lists the main classes and interfaces that JCA deals with. The list has been designed based on how useful and frequently used a class/interface is.

Table 4-2. *Classes/Interfaces and Their Purpose*

Class/Interface	Purpose
Provider	This generates a pair of public and private keys for a certain algorithm.
Security	For an algorithm, this class will generate new secret keys.
MessageDigest	An instance of this class is used by one or two parties for establishing a cryptographic key to be used within a specific cryptographic operation.
Cipher	This class will use the keys (generated for encryption and decryption data). The class provides a set of various algorithms, such as asymmetric encryption, symmetric encryption, and password-based encryption.
KeyAgreement	This provides functionalities for computing the hash value for a certain message.
CertificateFactory	This provides functionalities for generating random numbers.
Key	This provides functionalities for working with keys.

For a comprehensive list, the resource listed in [1] should be consulted.

The most important packages that are used in practice are java.security, javax.crypto, javax.crypto.interfaces, java.security.interfaces, javax.crypto.spec, java.security.cert, and java.security.spec.

Data Encryption

To work with data encryption, the Cipher class offers the right tools (e.g., functions, methods, etc.). The Cipher class is part of the javax.crypto package.

The workflow is straightforward to follow, as shown here and in Listing 4-1:

1. Declare a KeyPairGenerator object.

2. The KeyPairGenerator has to be initialized accordingly, specifying the algorithm that will be used.

3. Using the method generateKeyPair(), you will generate the key pair.

4. Obtain the public key using getPublic().

5. Create an instance for the cipher object. To achieve this goal, the method getInstance() is used.

6. Using the init() method, you will initialize the cipher object.

7. Provide the data to the cipher object for which you will provide the encryption.

8. Provide the encryption of the data.

Listing 4-1. Encryption Example

```
1   import java.security.KeyPair;
2   import java.security.KeyPairGenerator;
3   import java.security.Signature;
4
5   import javax.crypto.Cipher;
6
7   public class DataEncryption {
8       public static void main(String args[]) throws Exception{
9
10          //create an instance for signature
11          Signature mySignature = Signature.getInstance("SHA256withRSA");
12          System.out.println("\n The signature instance -> " +
             mySignature);
13
14          //crere an instance as generator for a pair key
15          KeyPairGenerator key_pair_gen = KeyPairGenerator.
             getInstance("RSA");
16          System.out.println("\n The key pair generator -> " +
             key_pair_gen);
17
18          //specify the size
19          key_pair_gen.initialize(2048);
```

```
20        System.out.println("\n The key pair generator size -> " +
          key_pair_gen);
21
22        //create the key pair instance
23        KeyPair pair = key_pair_gen.generateKeyPair();
24        System.out.println("\n The instance of the key pair -> "
          + pair);
25
26        //specify the algorithm as instance
27        Cipher algorithm = Cipher.getInstance("RSA/ECB/PKCS1Padding");
28        System.out.println("\n The algorithm -> " + algorithm);
29
30        //choose the mode (encryption/decryption)
31        algorithm.init(Cipher.ENCRYPT_MODE, pair.getPublic());
32        System.out.println("\n The algorithm with the mode -> " +
          algorithm);
33
34        //choose a text that will be encrypted
35        byte[] input = "Learning Cryptography with Java".getBytes();
36        algorithm.update(input);
37        System.out.println("\n The algorithm with the mode -> " +
          algorithm);
38
39        //encrypting the data
40        byte[] encrypted_text = algorithm.doFinal();
41        System.out.println("\n Encrypted text -> " + new String
          (encrypted_text, "UTF8"));
42    }
43 }
```

Figure 4-3 shows the output.

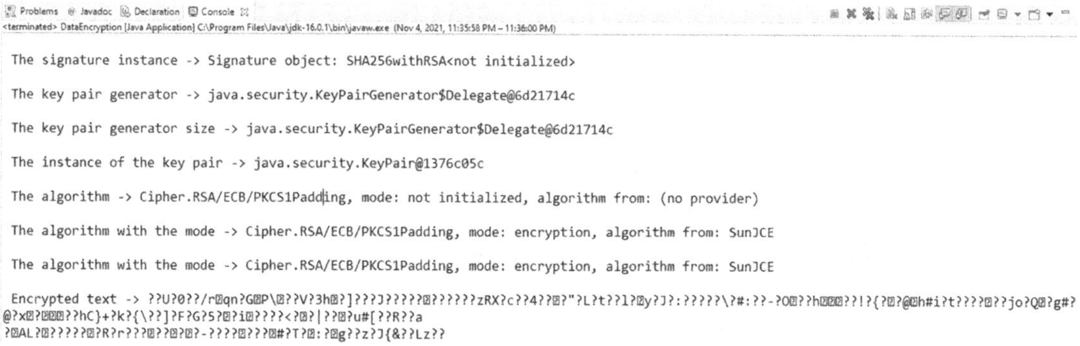

Figure 4-3. *The output of the encryption process*

Hash Functions

To use hash functions (Listing 4-2 and Figure 4-4), the process is simple and generally is composed of three steps, as follows:

1. Provide an instance for the MessageDigest class.

2. Initialize the MessageDigest instance with data.

3. Generate the hash of the provided message.

Listing 4-2. Generating Hash Functions

```
1  import java.security.MessageDigest;
2  import java.util.Scanner;
3
4  public class Hashing {
5     public static void main(String args[]) throws Exception
6     {
7        //add some text that we will use it for encryption
8        String message = "Welcome to Appress and enjoy learning
          cryptography";
9        System.out.println("\n The text for which we will provide the
          digest (hash) is -> " + message);
10
```

```
11          //declare an instance for which we will provide the hash digest
            using SHA-256
12          MessageDigest messsage_digest = MessageDigest.
            getInstance("SHA-256");
13
14          //get the bytes representation by providing data for updating
            the digest
15          messsage_digest.update(message.getBytes());
16
17          //obtain the hash of the message
18          byte[] digest = messsage_digest.digest();
19          System.out.println(digest);
20
21          //provide conversion from byte to hex string format
22          StringBuffer hex_representation = new StringBuffer();
23
24          for (int i = 0;i<digest.length;i++)
25          {
26              hex_representation.append(Integer.toHexString(0xFF &
                digest[i]));
27          }
28          System.out.println("The hex representation : " +
            hex_representation.toString());
29      }
30  }
```

```
[Problems] [Javadoc] [Declaration] [Console ⊠                                              ⬛ ✖ ❌ | 📭 🔛 🔛 🔛 | 🔛 ▢ ▾ ▢ ▾ ▾
<terminated> Hashing [Java Application] C:\Program Files\Java\jdk-16.0.1\bin\javaw.exe (Nov 4, 2021, 11:38:09 PM – 11:38:09 PM)

 The text for which we will provide the digest (hash) is -> Welcome to Appress and enjoy learning cryptography
[B@41cf53f9
The hex representation : b2601ff0a0ff94bbd0d5f398b5412ccbc471592b6a20be7a8f38893ae6f1144e
```

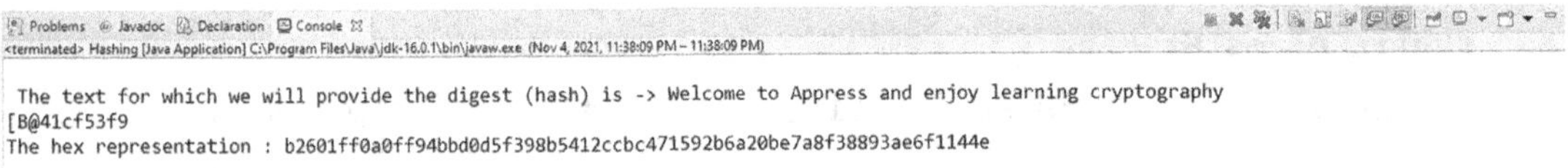

Figure 4-4. *The hash function output*

Signatures

Creating and working with signatures in Java represents an interesting approach compared with other programming languages. Digital signatures give you the ability to verify different aspects to make certain a message is genuine, such as the following:

- Verifying the author

- Checking the date and time of signatures

- Providing authentication of the message contents

Digital signatures offer the following advantages:

- *Authentication*: Digital signatures will help you properly authenticate the source that is sending the message(s).

- *Integrity*: The signature will be invalided if any change is made on the message.

- *Nonrepudiation*: Once a signature has been provided by someone or an entity, there is no chance at a certain point to deny the process of signing.

Generating Signatures

The implementation process of generating signatures involves eight steps, as follows:

1. Declare an instance of the `KeyPairGenerator` class.

2. Provide the proper initialization for the instance.

3. Obtain the key pair using the `generateKeyPair()` method.

4. Extract the private key from the pair using the `getPrivate()` method.

5. Declare an instance as a signature object.

6. Initialize the signature object.

7. Provide the data for the signature object.

8. Compute the signature.

Listing 4-3 shows how the steps listed are implemented, with an explanation of each step. Figure 4-5 shows the result.

Listing 4-3. Example for Generating a Signature

```java
1   import java.security.*;
2   import java.util.Scanner;
3
4   public class GeneratingSignature
5   {
6       public static void main(String args[]) throws Exception {
7           //the message that will be signed
8           String data_message = "Welcome to Apress and enjoy learning
            cryptography";
9           System.out.println("The message that will be sign is -> " +
            data_message);
10
11          //Step 1 - Declare an instance of KeyPairGenerator class
12          KeyPairGenerator generatorKeyPair = KeyPairGenerator.
            getInstance("DSA");
13
14          //Step 2 - Providing the proper initialization for the instance
15          generatorKeyPair.initialize(2048);
16
17          //Step 3 - Obtaining the key pair using the
            generateKeyPair() method
18          KeyPair cryptographyKeyPair = generatorKeyPair.
            generateKeyPair();
19
20          //Step 4. Extract the private key from the pair using the
            getPrivate() method
21          PrivateKey private_key = cryptographyKeyPair.getPrivate();
22
23          //Step 5. Declare an instance as a signature object
24          Signature signatureForMessage = Signature.getInstance
            ("SHA256withDSA");
```

```
25
26          //Step 6. Initialize the signature object
27          signatureForMessage.initSign(private_key);
28          byte[] theBytesRepresentation = "msg".getBytes();
29
30          //Step 7. Provide the data for the signature object
31          signatureForMessage.update(theBytesRepresentation);
32
33          //Step 8. Compute the signature
34          byte[] signature = signatureForMessage.sign();
35
36          //Showing the result as output
37          System.out.println("Obtained digital signature for the provided
            text ->  "+new String(signature, "UTF8"));
38      }
39  }
```

Problems Javadoc Declaration Console
<terminated> VerifySignature [Java Application] C:\Program Files\Java\jdk-16.0.1\bin\javaw.exe (Nov 5, 2021, 1:50:56 AM – 1:50:56 AM)
The message that will be sign is -> Welcome to Apress and enjoy learning cryptography
Obtained digital signature for the provided text -> 0<▯   mi▯?   5GBR?]M5&?.Q▯F????X?vg▯|?[??????▯??1??+)???^▯??

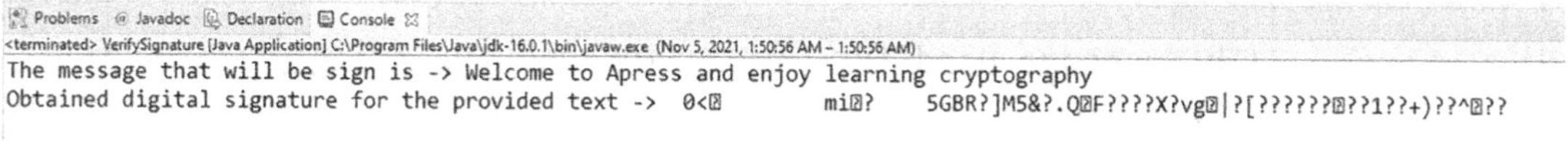

Figure 4-5. *Output for generating a signature*

Verifying the Signature

The verifying signature process is different from the generating process in terms of the
steps that you need to follow for the implementation. The steps are as shown here:

1. Declare an instance of the KeyPairGenerator class.

2. Provide the proper initialization for the instance.

3. Obtain the key pair using the generateKeyPair() method.

4. Extract the private key from the pair.

5. Declare an instance for the signature object.

6. Provide the proper initialization for the signature object.

7. Provide data for the signature instance.

8.   Compute the signature.

9.   Provide verification for the signature instance.

10. Provide verification for the data by updating the signature.

11. Invoke the verifying process for the signature.

Listing 4-4 shows how the steps are implemented, and it points out the main differences between the generating process and the verifying process. Figure 4-6 shows the results.

Listing 4-4. Implementation for generating and validating a signature

```
1    import java.security.*;
2    import java.util.Scanner;
3
4    public class VerifyingSignature
5    {
6       public static void main(String args[]) throws Exception
7       {
8               //The message used for generating and verifying process of
                the signature
9               String message_data = "Welcome to Apress and enjoy learning
                cryptography";
10              System.out.println("The message used -> " + message_data);
11
12              //Step 1. Declare an instance of KeyPairGenerator class
13              KeyPairGenerator generating_key_pair = KeyPairGenerator.
                getInstance("DSA");
14
15              //Step 2. Providing the proper initialization for the instance
16              generating_key_pair.initialize(2048);
17
18              //Step 3. Obtaining the key pair using the
                generateKeyPair() method
19              KeyPair cryptographic_pair = generating_key_pair.
                generateKeyPair();
20
```

44

```java
21          //Step 4. Extract the private key from the pair
22          PrivateKey private_crypto_key = cryptographic_pair.
            getPrivate();
23
24          //Step 5 - Declare an instance for the signature object
25          Signature signature = Signature.getInstance("SHA256withDSA");
26
27          //Step 6. Providing the proper initialization for the
            signature object
28          signature.initSign(private_crypto_key);
29          byte[] bytes = message_data.getBytes();
30          System.out.println("The representation as bytes is: " + bytes);
31
32          //Step 7. Provide data for the signature instance
33          signature.update(bytes);
34
35          //Step 8. Compute the signature
36          byte[] computedSignature = signature.sign();
37
38          //Step 9. Provide verification for the signature instance
39          signature.initVerify(cryptographic_pair.getPublic());
40          signature.update(bytes);
41
42          //Step 10. Provide verification for the data through updating
            the signature, and
43          //Step 11. Invoke the verifying process for the signature
44          boolean bool = signature.verify(computedSignature);
45
46          if(bool)
47          {
48              System.out.println("The signature has been checked
                (verified) with success");
49          }
```

```
50              else
51          {
52                  System.out.println("The signature has been failed");
53          }
54      }
55  }
```

Problems   @ Javadoc   Declaration   Console ⋈
`<terminated> VerifyingSignature [Java Application] C:\Program Files\Java\jdk-16.0.1\bin\javaw.exe (Nov 5, 2021, 2:28:46 AM – 2:28:48 AM`

```
The message used -> Welcome to Apress and enjoy learning cryptography
The representation as bytes is: [B@b1a58a3
The signature has been checked (verified) with success
```

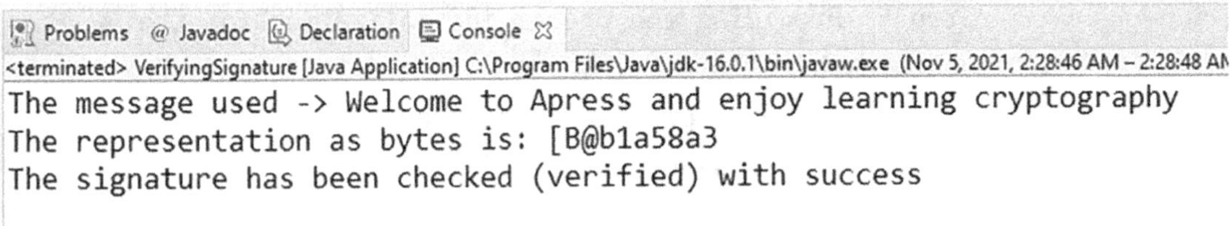

Figure 4-6. *The verifying process*

Conclusion

This chapter covered some of the main classes specific to the Java Cryptography Architecture. The chapter explained the basic concepts of JCA, and how understanding how the JCA architecture was designed plays an important role in developing professional software applications.

You should now understand the basic process for implementing regular cryptography operations using JCA, including the following:

- Data encryption

- Hash functions

- Digital signatures (generating and verifying process)

References

[1].   Java Cryptography Architecture (JCA) Reference Guide, `https://docs.oracle.com/javase/8/docs/technotes/guides/security/crypto/CryptoSpec.html#Design`

[2].   Java Downloads, `https://www.oracle.com/java/technologies/downloads/#java17`

CHAPTER 5

Classical Cryptography

Humans have been hiding messages from the very first moment they started sending messages. In the early ages, messages were sent between leaders, emperors, conquerors, kings, and so on, in the form of verbal messages through a human messenger. Although unspoken rules protected the human messengers, sometimes this rule was not followed, and the messenger was caught by enemies. At that point, all the enemy needed to do was send an altered message to the rightful receiver. This is a real-life example of a *man-in-the-middle* attack.

In the years that followed, humans started hiding the actual content of the message in specific ways so that if the enemy took possession of the message, its meaning could not be understood. In this way, an early form of cryptography was born.

Methods of encrypting messages are as limitless as the human imagination. However, in practice, only some of them work well due to some natural constraints. For example, the sender and the receiver should both know how to encrypt and decrypt the messages, and the encryption technique should be strong enough such that if the encrypted message gets into the wrong hands, it cannot be decrypted. These two simple facts haunt all encryption schemes even nowadays. In this chapter, some historical ciphers will be presented, as follows: Caesar cipher, Vigenère cipher, and Hill cipher.

All these classical ciphers fall into the category of *symmetric cryptography*, in which just one type of key is used for both encryption and decryption, namely, the *secret key* (we'll discuss symmetric cryptography more in Chapter 9). Although all these classical ciphers are not used anymore, they are important as an introduction to cryptography so you can understand the basic concepts and the purpose behind cryptography. Also, they are easy to understand from a technical perspective, which makes them again a great

© Stefania Loredana Nita and Marius Iulian Mihailescu 2022
S. L. Nita and M. I. Mihailescu, *Cryptography and Cryptanalysis in Java*,
https://doi.org/10.1007/978-1-4842-8105-5_5

introduction to cryptography, especially for beginners in this field. You can find more background information about the ciphers in this chapter in many cryptography courses and books, such as [1], [2], and [3].

Caesar Cipher

The Caesar cipher is estimated to date back to the first century BC in the Roman Empire. Its name comes from the Roman Emperor Julius Caesar, who used this cipher to encrypt military messages and strategies [1], and it is estimated to be broken in about the fifth century AD. The Caesar cipher is a substitution cipher, in which each letter of the alphabet is moved a certain number of characters to the right. For example, if the established number is 5, then *A* will become *F*, *B* will become *G*, *C* will become *H*, etc. In this cipher, the *key* is represented by the number that the letters are shifted. Note that the key should be secret, known only by the sender (to encrypt the messages) and by the receiver (to decrypt the messages). The Caesar cipher is important because it is the basis for other classical ciphers, such as Vigenère. The Caesar cipher with key $k = 13$ is also known as Rot13.

Let's consider an example. The key for the cipher is set to $k = 6$; therefore, all letters are shifted six characters (see Table 5-1).

Table 5-1. *Caesar Representation with k = 6*

0	1	2	3	4	5	6	7	8	9	10	11	12	13	14	15	16	17	18	19	20	21	22	23	24	25
A	B	C	D	E	F	G	H	I	J	K	L	M	N	O	P	Q	R	S	T	U	V	W	X	Y	Z
G	H	I	J	K	L	M	N	O	P	Q	R	S	T	U	V	W	X	Y	Z	A	B	C	D	E	F

Note that there are 26 letters in the alphabet, and in Table 5-1 the numbering starts with 0 and ends with 25 because it is related to $\mathbb{Z}_{26} = \{0, 1, ..., 25\}$. Sometimes, in other documentation, the numbering starts with 1 and ends with 26, because it seems more natural.

Considering the plain message as "CAESAR CIPHER EXAMPLE" and the key as $k = 6$, the encrypted message will be "IGKYGX IOVNKX KDGSVRK." The encrypted message was obtained by replacing *C* with the corresponding *I*, *A* with the corresponding *G*, *E* with the corresponding *K*, and so on.

To decrypt the message, the letters from the second row of the table are shifted to the left for k positions. Therefore, the decryption of "IGKYGX IOVNKX KDGSVRK" (with the key $k = 6$) is "CAESAR CIPHER EXAMPLE," obtained by applying a similar logic as for encryption: I is replaced with C, G is replaced with A, and so on. Note that sometimes the space character is removed from the message; therefore, in the example, the plain text would be "CAESARCIPHEREXAMPLE," while the encrypted text would be "IGKYGXIOVNKXKDGSVRK."

From a mathematical point of view, the operations are addition for encryption and subtraction for decryption via a modulo operation. The first step is to "convert" the characters to numbers, as follows: $A \rightarrow 1, B \rightarrow 2, ..., Z \rightarrow 26$. The encryption function enc_k will take as input one parameter, i.e., the number corresponding to the letter that needs to be encrypted at the current step, and will output the encrypted letter:

$$enc_k(c) = (c + k) \bmod 26,$$

where c is the plain character and k is the key. The decryption function needs to shift back the letters and, therefore, has the following representation:

$$dec_k(c') = (c' - k) \bmod 26,$$

where c' is the encrypted character that needs to be decrypted and k is the key.

Implementation

Listing 5-1 shows the implementation of the Caesar cipher; Figure 5-1 shows the result.

Listing 5-1. Implementation of the Caesar Cipher

```
1   import java.util.Scanner;
2
3   public class CaesarCipher {
4
5       public static String Encrypt(String text, int positions) {
6           String toEncrypt = "", result = "";
7           //format the text to be encrypted
8           for (int i = 0; i < text.length(); i++) {
9               //remove space characters
```

```
10                        if (text.charAt(i) == ' ')
11                              continue;
12                        else {
13                              //if the character is lowercase, make it
                                uppercase
14                              if (Character.isLowerCase(text.charAt(i)))
15                                    toEncrypt += Character.toUpperCase
                                        (text.charAt(i));
16                              //otherwise keep the uppercase character
17                              else
18                                    toEncrypt += text.charAt(i);
19                        }
20                  }
21            for (int i = 0; i < toEncrypt.length(); i++) {
22                  //shift the current letter of the message with given
                      positions to right
23                  char shiftedLetter = (char) (toEncrypt.charAt(i) +
                      positions);
24
25                  //if the ASCII code exceeds Z, then bring it back in
                      the interval A..Z
26                  if (shiftedLetter > 'Z')
27                        shiftedLetter = (char) (shiftedLetter +
                            'A' - 'Z' - 1);
28
29                  result += shiftedLetter;
30            }
31
32        return result;
33      }
34
35      public static String Decrypt(String text, int positions) {
36            //the encrypted code is already uppercase,
37            //therefore there is no need of formatting
38            String result = "";
```

```
39          for (int i = 0; i < text.length(); i++) {
40
41                  //shift the current letter of the message with given
                    positions to left
42                  char shiftedLetter = (char) (text.charAt(i) -
                    positions);
43
44                  //if the ASCII code exceeds A, then bring it back in
                    the interval A..Z
45                  if (shiftedLetter < 'A')
46                          shiftedLetter = (char) (shiftedLetter -
                        'A' + 'Z' + 1);
47
48                  result += shiftedLetter;
49
50          }
51
52      return result;
53  }
54
55  public static void main(String[] args) {
56          System.out.println("CAESAR CIPHER\n");
57
58          Scanner sc = new Scanner(System.in);
59
60          // Reading the input: plain message and the secret key
61          System.out.print("Type the message: ");
62          String message = sc.nextLine();
63          System.out.print("Type the key: ");
64          int key = sc.nextInt();
65          sc.close();
66
67          // Encrypting the plain message
68          System.out.println("\nEncrypting...");
69          String encryptedMessage = Encrypt(message, key);
```

```
70              System.out.println("The encrypted message is: " +
                encryptedMessage);
71
72              // Decrypting the encrypted message
73              System.out.println("\nDecrypting...");
74              String recoveredMessage = Decrypt(encryptedMessage, key);
75              System.out.println("The decrypted message is: " +
                recoveredMessage);
76
77          }
78      }
```

The code is pretty simple and just follows the encryption and decryption formula. Before encrypting the message, it is formatted by eliminating the space characters and making all the letters uppercase.

Figure 5-1. *The result of Caesar enciphering*

Cryptanalysis

The Caesar cipher is easy to use, but it is also easily broken in milliseconds using automated tools [5]. The total number of keys is 25; therefore, there are only 25 shifting possibilities. An attacker can easily generate all 25 forms and then recover the message for all forms. This type of attack, in which all possible values are exploited, is called the *brute-force attack.*

The brute-force attack for the Caesar cipher consists of determining the fitness value of a part of the decrypted text, showing how that part is fitting in the text. This implies some statistical information about the decrypted text compared to statistical information about English text. The fitness value shows how "close" to recover the key an attacker is, with higher values closer to recovering the key. An example of a statistical method that can be applied here is the quadgram fitness measure [4] that works as follows:

1. Compute the static values for the English text.

2. Compute the probability for the event that the ciphertext has the same distribution.

Vigenère Cipher

This historical cipher was mistakenly attributed to Blaise de Vigenère in the 18th century, but it was invented by the Italian cryptologist Giovan Battista Bellaso (https://en.wikipedia.org/wiki/Giovan_Battista_Bellaso), and the first use is estimated to be in the 15th century. Until 1863, when it was broken through a method proposed by Friedrich Kasiski, the Vigenère cipher was considered unbreakable, gaining the nickname "le chiffrage indéchiffrable" (in French), meaning "the indecipherable cipher."

The Vigenère cipher uses a key in a form of a word and an encryption table (see Table 5-2). To encrypt a message, it should proceed as follows:

1. Extend the key until it reaches the length of the message, by repeating circularly the letters of the key.

2. For each letter of the message, find the intersection in the encryption table between the current letter and the letter with the same index in the extended key.

3. The found letter is the encrypted letter of the current plain letter.

To make it easier to understand, see the example in Table 5-2.

Table 5-2. *The Encryption Table for the Vigenère Cipher [6]*

```
    A B C D E F G H I J K L M N O P Q R S T U V W X Y Z
    -------------------------------------------------------
A   A B C D E F G H I J K L M N O P Q R S T U V W X Y Z
B   B C D E F G H I J K L M N O P Q R S T U V W X Y Z A
C   C D E F G H I J K L M N O P Q R S T U V W X Y Z A B
D   D E F G H I J K L M N O P Q R S T U V W X Y Z A B C
E   E F G H I J K L M N O P Q R S T U V W X Y Z A B C D
F   F G H I J K L M N O P Q R S T U V W X Y Z A B C D E
G   G H I J K L M N O P Q R S T U V W X Y Z A B C D E F
H   H I J K L M N O P Q R S T U V W X Y Z A B C D E F G
I   I J K L M N O P Q R S T U V W X Y Z A B C D E F G H
J   J K L M N O P Q R S T U V W X Y Z A B C D E F G H I
K   K L M N O P Q R S T U V W X Y Z A B C D E F G H I J
L   L M N O P Q R S T U V W X Y Z A B C D E F G H I J K
M   M N O P Q R S T U V W X Y Z A B C D E F G H I J K L
N   N O P Q R S T U V W X Y Z A B C D E F G H I J K L M
O   O P Q R S T U V W X Y Z A B C D E F G H I J K L M N
P   P Q R S T U V W X Y Z A B C D E F G H I J K L M N O
Q   Q R S T U V W X Y Z A B C D E F G H I J K L M N O P
R   R S T U V W X Y Z A B C D E F G H I J K L M N O P Q
S   S T U V W X Y Z A B C D E F G H I J K L M N O P Q R
T   T U V W X Y Z A B C D E F G H I J K L M N O P Q R S
U   U V W X Y Z A B C D E F G H I J K L M N O P Q R S T
V   V W X Y Z A B C D E F G H I J K L M N O P Q R S T U
W   W X Y Z A B C D E F G H I J K L M N O P Q R S T U V
X   X Y Z A B C D E F G H I J K L M N O P Q R S T U V W
Y   Y Z A B C D E F G H I J K L M N O P Q R S T U V W X
Z   Z A B C D E F G H I J K L M N O P Q R S T U V W X Y
```

Consider the plaintext "LOREM IPSUM DOLOR SIT AMET" and the key "APRESS."

First, extend the key, as in Table 5-3. Note that the space character was removed from the plain message.

Table 5-3. *Extending the Key for Vigenère Enciphering*

	1	2	3	4	5	6	7	8	9	10	11	12	13	14	15	16	17	18	19	20	21	22
Text	L	O	R	E	M	I	P	S	U	M	D	O	L	O	R	S	I	T	A	M	E	T
Key	A	P	R	E	S	S	A	P	R	E	S	S	A	P	R	E	S	S	A	P	R	E

Second, find the intersection between the current key letter and the current message letter (see Figure 5-2).

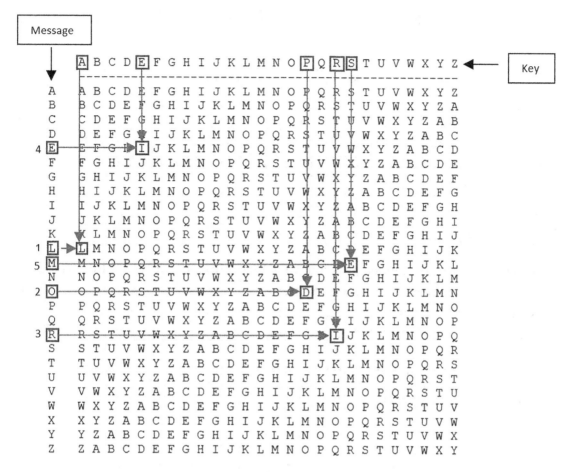

Figure 5-2. *Encryption of the first word of the message using Vigenère*

Continuing with the encryption technique, the encrypted message is "LDIIEAPHLQVGLDIWALABVX."

The Vigenère cipher is a substitution cipher in which each letter of the plain message is replaced with the letter of the intersection between the key and the plain message, as shown earlier.

The technical aspects of the Vigenère cipher are related to $\mathbb{Z}_{26} = \{0, 1, ..., 25\}$. Considering m_i the current letter of the message and k_i the current letter of the extended key, the corresponding encrypted letter is computed using the following formula:

$$enc(m_i) = (m_i + k_i) \bmod 26$$

The decryption of the current letter c_i of the encrypted message is made using the following formula:

$$dec(c_i) = (c_i - k_i) \bmod 26,$$

where k_i is the current letter of the extended key.

The Vigenère cipher is important because it was the starting point of the one-time pad (OTP) proposed by Gilbert Vernam in the 1900s, which is the basis for authentication systems today.

Implementation

This section presents the implementation of the Vigenère cipher. It follows the steps described in the previous section, but first, the message is processed by eliminating the space characters and making the letters uppercase. Listing 5-2 shows the implementation, and Figure 5-3 shows the result.

Listing 5-2. Implementation of the Vigenère Cipher

```
1    import java.util.Scanner;
2
3    public class Vigenere {
4         static String ExtendKey(String message, String initialKey) {
5             int toLength = message.length();
6
7             for (int i = 0;; i++) {
8                 if (toLength == i)
9                     i = 0;
10                if (initialKey.length() == message.length())
11                    break;
12                initialKey += initialKey.charAt(i);
13            }
14            return initialKey;
15        }
16
```

```
17      static String Encrypt(String text, String extendedKey) {
18          String result = "";
19
20          for (int i = 0; i < text.length(); i++) {
21              int toInt = (text.charAt(i) + extendedKey.
                charAt(i)) % 26;
22              toInt += 'A';
23
24              result += (char) (toInt);
25          }
26          return result;
27      }
28
29      static String Decrypt(String text, String extendedKey) {
30          String result = "";
31
32          for (int i = 0; i < text.length() && i < extendedKey.
            length(); i++) {
33
34              int toInt = (text.charAt(i) - extendedKey.charAt(i)
                + 26) % 26;
35              toInt += 'A';
36              result += (char) (toInt);
37          }
38          return result;
39      }
40
41      static String ProcessText(String text) {
42          String result = "";
43
44          for (int i = 0; i < text.length(); i++) {
45              if (text.charAt(i) == ' ')
46                  continue;
47              else {
48                  if (Character.isLowerCase(text.charAt(i)))
```

```
49                                          result += Character.toUpperCase(text.
                                            charAt(i));
50                              else
51                                          result += text.charAt(i);
52                  }
53              }
54
55          return result;
56      }
57
58      public static void main(String[] args) {
59              System.out.println("VIGENERE CIPHER\n");
60
61              Scanner sc = new Scanner(System.in);
62
63              System.out.print("Type the message: ");
64              String plainMessage = sc.nextLine();
65              System.out.print("Type the key: ");
66              String key = sc.nextLine();
67              sc.close();
68
69              System.out.println("\nEncrypting...");
70
71              String processedText = ProcessText(plainMessage);
72              String keyword = ProcessText(key);
73
74              String extendedKey = ExtendKey(processedText, keyword);
75              String encryptedText = Encrypt(processedText,
                extendedKey);
76
77              System.out.println("The encrypted message is: " +
                encryptedText);
78              System.out.println("\nDecrypting...");
79              String recoveredMessage = Decrypt(encryptedText,
                extendedKey);
```

```
80          System.out.println("The decrypted message is: " +
                recoveredMessage);
81        }
82    }
```

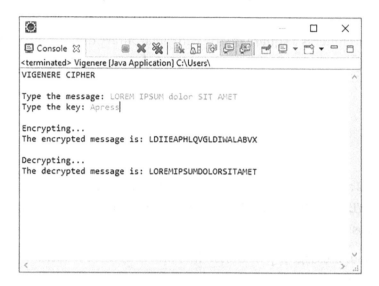

Figure 5-3. *The result of the Vigenère encryption*

Cryptanalysis

The Vigenère cipher can be broken using frequency analysis. However, it is not enough to do a simple analysis frequency because the same letter can be obtained from intersections of the plain message letter and the extended key letter. It would be easier if the length of the original key was discovered; let's denote it with n, because the encrypted text will be partitioned in blocks of length n (except for the last block that may have a smaller length). From here, the problem is reduced to breaking n Caesar ciphers. To find the key length in general, three techniques can be applied: brute-force, Kasiski examination [7], or Friedman test [8].

Hill Cipher

The author of the Hill cipher is Lester S. Hill, and it was proposed in 1929; it is an example of a polygraphic substitution cipher. The mathematical background for the Hill cipher lies in linear algebra, more exactly in matrices and their operations, in which the plaintext is hidden.

The key of the cipher is a square matrix with an arbitrary dimension, with elements in $\mathbb{Z}_{26}$. For example, consider the following matrix to be the secret key,

$$K = \begin{bmatrix} 2 & 8 & 5 \\ 9 & 4 & 1 \\ 6 & 17 & 7 \end{bmatrix},$$

having three rows and three columns (denote the dimension with $n = 3$), and the plain message to be "WELCOME TO CRYPTOGRAPHY." To encrypt the message, it should proceed as follows:

1. The plain message is partitioned into blocks of three letters (because $n = 3$).

2. Convert the letters in corresponding indices of the alphabet (for convenience, see the first two rows of Table 5-1).

3. Multiply the key (square matrix) with the column matrix that represents the indices of the letters and apply modulo 26.

4. Convert the numbers into letters (for convenience, see the first two rows of Table 5-1).

Following the example, for the first three letters of the message, it will be obtained $'W' = 22$, $'E' = 4$, and $'L' = 11$, and then

$$\begin{bmatrix} 2 & 8 & 5 \\ 9 & 4 & 1 \\ 6 & 17 & 7 \end{bmatrix} \begin{bmatrix} 22 \\ 4 \\ 11 \end{bmatrix} (mod\ 26) = \begin{bmatrix} 131 \\ 225 \\ 277 \end{bmatrix} (mod\ 26) = \begin{bmatrix} 1 \\ 17 \\ 17 \end{bmatrix} \rightarrow BRR$$

Therefore, the corresponding encrypted message of "WEL" is "BRR." The process will continue until the entire message is encrypted, and for the previous example we will obtain "BRRUIWWWDAGBSRRSSNYFN." For the cases in which the last block has fewer letters than n, some extra letters will be added to the end of the plain message to achieve n letters. Note that not any matrix can be used as a key. To be invertible modulo 26, the determinant of the matrix should be coprime with 26.

To decrypt the message, the first step is to find the inverse of K (let's denote it with K') with elements also in $\mathbb{Z}_{26}$; the next steps is to multiply K' with the column matrix representing the encrypted blocks. The column matrix that will be obtained represents the indices of the letters of the plain message.

$$K'(mod\ 26) = \begin{bmatrix} 21 & 1 & 22 \\ 7 & 12 & 23 \\ 17 & 22 & 22 \end{bmatrix}$$

$$\begin{bmatrix} 21 & 1 & 22 \\ 7 & 12 & 23 \\ 17 & 22 & 22 \end{bmatrix}\begin{bmatrix} 1 \\ 17 \\ 17 \end{bmatrix}(mod\ 26) = \begin{bmatrix} 412 \\ 602 \\ 765 \end{bmatrix}(mod\ 26) = \begin{bmatrix} 22 \\ 4 \\ 11 \end{bmatrix} \rightarrow WEL$$

Implementation

Listing 5-3 shows the implementation of the Hill cipher, and Figure 5-4 shows the result.

Listing 5-3. Hill Cipher Implementation

```
1    package ciphers;
2
3    import java.util.*;
4
5    class Utils {
6            String alphabet = "ABCDEFGHIJKLMNOPQRSTUVWXYZ";
7
8            int getCharIndex(char c) {
9                    if (alphabet.contains(Character.toString(c)))
10                            return alphabet.indexOf(c);
11                   else
12                            return -1;
13            }
14
15            char getCharAtIndex(int index) {
```

```
16                    return alphabet.charAt(index);
17            }
18    }
19
20    public class HillCipher {
21
22            Utils util = new Utils();
23            int keySize = 3;
24            int keyMatrix[][] = new int[keySize][keySize];
25            Scanner scanner = new Scanner(System.in);
26
27            HillCipher(int keySize) {
28                    this.keySize = keySize;
29            }
30
31            void keyInitialize(String message) throws Exception {
32                    System.out.println(message);
33                    for (int i = 0; i < keySize; i++) {
34                            for (int j = 0; j < keySize; j++) {
35                                    keyMatrix[i][j] = scanner.nextInt();
36                            }
37                    }
38            }
39
40            String blockEncryption(String plainBlock) throws Exception {
41                    plainBlock = plainBlock.toUpperCase();
42                    int indexMatrix[][] = new int[keySize][1];
43                    int columnMatrix[][] = new int[keySize][1];
44                    int sumElem = 0;
45                    String encryptedBlock = "";
46
47                    for (int i = 0; i < keySize; i++) {
48                            indexMatrix[i][0] = util.getCharIndex(plainBlock.
                            charAt(i));
49                    }
50
```

```
51                  for (int i = 0; i < keySize; i++) {
52                      for (int j = 0; j < 1; j++) {
53                          for (int k = 0; k < keySize; k++) {
54                              sumElem = sumElem + keyMatrix[i][k] *
                                 indexMatrix[k][j];
55                          }
56                          columnMatrix[i][j] = sumElem % 26;
57                          sumElem = 0;
58                      }
59                  }
60
61              for (int i = 0; i < keySize; i++) {
62                  encryptedBlock += util.
                       getCharAtIndex(columnMatrix[i][0]);
63              }
64              return encryptedBlock;
65          }
66
67      String messageEncryption(String plainMessage) throws
        Exception {
68              String encryptedMessage = "";
69              keyInitialize("Enter the values for key matrix:");
70
71              System.out.println("\n[Encrypting...]\n");
72
73              plainMessage = plainMessage.toUpperCase();
74
75              int messageLen = plainMessage.length();
76
77              while (messageLen % keySize != 0) {
78                  plainMessage += "Z";
79                  System.out.println(messageLen);
80                  messageLen = plainMessage.length();
81              }
82
```

```
83              for (int i = 0; i < messageLen - 1; i = i + keySize) {
84                      encryptedMessage += blockEncryption(plainMessage.
                        substring(i, i + keySize));
85              }
86              return encryptedMessage;
87          }
88
89      String blockDecryption(String encryptedBlock) throws
        Exception {
90              encryptedBlock = encryptedBlock.toUpperCase();
91              int indexMatrix[][] = new int[keySize][1];
92              int columnMatrix[][] = new int[keySize][1];
93              int sumElem = 0;
94              String decryptedBlock = "";
95
96              for (int i = 0; i < keySize; i++) {
97                      indexMatrix[i][0] = util.
                        getCharIndex(encryptedBlock.charAt(i));
98              }
99
100             for (int i = 0; i < keySize; i++) {
101                     for (int j = 0; j < 1; j++) {
102                         for (int k = 0; k < keySize; k++) {
103                                 sumElem = sumElem + keyMatrix[i][k] *
                                    indexMatrix[k][j];
104                         }
105                         while (sumElem < 0) {
106                                 sumElem += 26;
107                         }
108                         columnMatrix[i][j] = sumElem % 26;
109                         sumElem = 0;
110                     }
111             }
112
```

```
113             for (int i = 0; i < keySize; i++) {
114                 decryptedBlock += util.
                        getCharAtIndex(columnMatrix[i][0]);
115             }
116             return decryptedBlock;
117         }
118
119     String messageDecryption(String encryptedMessage) throws
        Exception {
120             String decryptedMessage = "";
121             keyInitialize("\n---\n\n Enter the values for inverted
                key matrix:");
122
123             System.out.println("\n[Decrypting...]\n");
124
125             encryptedMessage = encryptedMessage.replaceAll(" ", "");
126
127             encryptedMessage = encryptedMessage.toUpperCase();
128
129             int messageLen = encryptedMessage.length();
130
131             for (int i = 0; i < messageLen - 1; i = i + keySize) {
132                 decryptedMessage += blockDecryption(encrypted
                        Message.substring(i, i + keySize));
133             }
134             return decryptedMessage;
135         }
136
137
138
139     public static void main(String[] args) throws Exception {
140
141             Scanner scanner = new Scanner(System.in);
142
```

```
143            String plainMessage;
144            System.out.println("Enter the plain message:");
145            plainMessage = scanner.nextLine();
146
147            int keySize;
148            System.out.println("Enter the size of the matrix key:");
149            keySize = scanner.nextInt();
150
151            HillCipher cipher = new HillCipher(keySize);
152
153            plainMessage = plainMessage.replaceAll(" ", "");
154
155            String encryptedMessage;
156            encryptedMessage = cipher.messageEncryption(plainMessage);
157
158            System.out.println("Obtained encryptd message: \n" +
                   encryptedMessage);
159
160
161            String decryptedMessage = cipher.messageDecryption
                   (encryptedMessage);
162            System.out.println("Obtained decrypted message:\n" +
                   decryptedMessage);
163
164        }
165
166    }
```

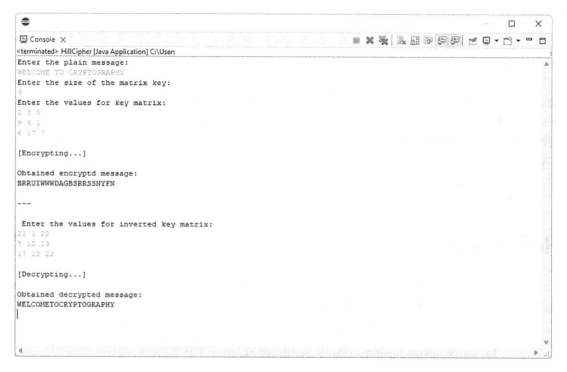

Figure 5-4. *The result of Listing 5-3*

The previous code follows the steps of the Hill cipher implementation. First, the plain message/encrypted message is split into blocks of the key's length. Then, each block is encrypted and added to decrypted message/plain message.

Other good references for Java implementations of the Hill cipher are [10] and [11].

Cryptanalysis

A known-plaintext attack can easily break the Hill cipher; therefore, if the attacker knows the plaintext and the corresponding encrypted message, the key can be recovered immediately, since all computations are linear. For a 2 × 2 key, the attack consists of computing the frequency for each digraph that may be obtained in the encrypted text. To launch this attack, knowing the standard English digraph representations is mandatory. For example, the group of letters *th* and *he* are common. A good resource for Hill cipher cryptanalysis is [9].

Conclusion

This chapter showed that primordial forms of cryptography existed in ancient times and presented three of the most important historical ciphers. These are considered symmetric ciphers because the same key is used for both encryption and decryption and it should be secret, known only by the sender and the receiver. Caesar, Vigenère, and Hill are simple ciphers from the technical perspective and are a good starting point for understanding cryptography. However, these ciphers are not used in practice because are easily cracked.

References

[1].   Menezes, A. J., Van Oorschot, P. C., & Vanstone, S. A. (2018). Handbook of applied cryptography. CRC press.

[2].   Paar, C., & Pelzl, J. (2009). Understanding cryptography: a textbook for students and practitioners. Springer Science & Business Media.

[3].   Stanoyevitch, A. (2010). Introduction to Cryptography with mathematical foundations and computer implementations. CRC Press.

[4].   Practical Cryptography (website). Quadgram Statistics as a Fitness Measure, http://practicalcryptography.com/cryptanalysis/text-characterisation/quadgrams/

[5].   Practical Cryptography (website). Cryptanalysis of Caesar Cipher, http://practicalcryptography.com/cryptanalysis/stochastic-searching/cryptanalysis-caesar-cipher/.

[6].   Mihailescu, M. I., & Nita, S. L. (2021). Cryptography and Cryptanalysis in MATLAB. Apress, Berkeley, CA.

[7].   Kasiski, F. W. (1863). Die Geheimschriften und die Dechiffrir-Kunst: mit besonderer Berücksichtigung der deutschen und der französischen Sprache. ES Mittler und sohn.

[8]. Van Tilborg, H. C., & Jajodia, S. (Eds.). (2014). Encyclopedia of cryptography and security. Springer Science & Business Media.

[9]. Practical Cryptography (website). Hill Cipher, `http://practicalcryptography.com/ciphers/classical-era/hill/`

[10]. Hill Cipher Program in Java, `https://www.javatpoint.com/hill-cipher-program-in-java`

[11]. Hill-Cipher, `https://github.com/tstevens/Hill-Cipher/blob/master/Main.java`

CHAPTER 6

Formal Techniques for Cryptography

This chapter discusses mathematical elements used in cryptography and throughout the remainder of this book. We'll start with some definitions of specific terminology; then, we'll cover some minimal notions of algebra; and finally, we'll delve into elliptic curves. This chapter contains a compilation of different technical (mathematical) tools used in cryptography gathered from different resources [1]–[9].

Definitions

Here are some definitions you need to know:

- *Sets*: The usual sets of numbers are $\mathbb{N}$ (natural numbers set), $\mathbb{Z}$ (integer numbers set), $\mathbb{R}$ (real numbers set), and $\mathbb{C}$ (complex numbers set). Other important sets used in cryptography are $\mathbb{Z}_n = \dfrac{\mathbb{Z}}{n\mathbb{Z}} = \{0, \ldots, n-1\}, n \geq 2$ (set of integers modulo integer n). Given a prime number p, the field $\mathbb{F}_p = \mathbb{Z}_p$. Lastly, $\mathcal{P}(S)$ contains the set of all subsets of S (where S is itself a set).

- *Bits*: An element of the set $\mathbb{Z}_2 = \{0, 1\}$ is called a bit. A *bit string* of length n is a vector that contains n bits. The notation $\{0, 1\}^n$ represents the set of bit strings of length n. To represent the set of all bit strings of finite length (including the empty string), the notation $\{0, 1\}^*$ is used.

- *Cryptographic primitive*: This represents "easier" algorithms used to construct cryptographic protocols or other complex cryptographic mechanisms. For example, the one-way hash function is a cryptographic primitive.

© Stefania Loredana Nita and Marius Iulian Mihailescu 2022
S. L. Nita and M. I. Mihailescu, *Cryptography and Cryptanalysis in Java*,
https://doi.org/10.1007/978-1-4842-8105-5_6

- *Cryptosystem*: This is a tuple of the form
  $\epsilon = (KeyGeneration,\ Encryption,\ Decryption)$ that consists of three
  probabilistic polynomial-time algorithms, i.e., the key generation
  algorithm, the encryption algorithm, and the decryption algorithm,
  that must fulfill the following condition:

$$Dec_{K'}\left(Enc_K\left(m\right)\right)=m,$$

  where K is the encryption key, K' is the decryption key, m is the plain
  message, *Enc* is the encryption algorithm, and *Dec* is the decryption
  algorithm. Note that according to the type of encryption, K may be
  the same as K', in which case it should be kept private, or K may be
  different from K', in which case K is public and K' is secret. You can
  find more about cryptosystems types in Chapter 9 and Chapter 10.

- *Hardness assumption*: This represents a computational assumption
  that states a problem cannot be solved efficiently (by efficiently it
  means in polynomial time regarding a quantity). Some common
  examples of hardness assumptions are integer factorization, discrete
  logarithm, cryptographic pairings, Diffie-Hellmann assumptions, etc.

- *Protocol.*: Usually, a protocol is applied between n entities and is a
  procedure with a set of rules through which messages are exchanged
  between the n entities. To construct messages, public sets and/or
  private sets are used based on cryptographic primitives. For message
  exchange, it can be required sequential communication.

- *Authorities and third parties*: Authorities are usually used to issue digital
  certificates, while trusted third parties are intermediary entities for
  reciprocal actions between two entities in which they both trust each other.

- *Attacks*: In cryptography, an attack is a technique that puts in danger the
  security of a system by searching for security weaknesses of the system and
  exploiting them for malicious purposes. The same definition applies to
  cryptanalysis, but the major difference between an attack and cryptanalysis
  is its purpose. The goal of cryptanalysis is to find the security weaknesses
  of a system, but the goal is to overcome those security issues and to provide
  improvements or security alternatives to protect the system.

Probabilities and Statistics

In cryptography, statistics and probabilities have an important role, especially in the security analysis process for cryptosystems, and in a branch of cryptography called *steganalysis*. This section covers basic statistics and probabilities notions. For more advanced concepts, you can consult [9]–[11].

Let's start by defining the notion of experiment. An *experiment* is a procedure through which a value is obtained from a predefined set of values $S = \{s_1, ..., s_n\}$, called a *sample space*. A sample is an individual value within the given set S. A specific value from S is called an *event*. A *discrete* sample space is a set for which the number of possible outcomes is limited.

A probability distribution over the sample space S is a set $K = \{k_1, ..., k_m\}$, with $K \subset S$, for which $k_1 + k_2 + ... + k_m = 1$. The value k_i is the probability that the event s_i will occur.

The probability of an event E is denoted with $P(E)$. When the event E contains values from S, $E = \{e_1, ..., e_k\}$, the probability of the event E is computed as $P(E) = P(e_1) + ... + P(e_k)$. The complementary event of E is denoted with $\bar{E}$ and means all values that do not belong to E.

The following relations occur between S, E, and $\bar{E}$, where $E \subset S$:

1. $0 \leq P(E) \leq 1$.

2. $P(E) = 0$ if $E = \phi$ (empty set); $P(E) = 1$ if $E = S$.

3. If every element of S has the same probability of occurring, then
$$P(E) = \frac{\#E}{\#S}$$ (where # is the cardinality of the set that follows the symbol).

4. $P(\bar{E}) = 1 - P(E)$.

Two events, E_1, E_2, are called mutually exclusive when $P(E_1 \cap E_2) = 0$.

Conditional Probability

Conditional probability occurs in situations in which one event depends on one or more other events. Let E_1 and E_2 be two events with $P(E_2) > 0$. The conditional probability for E_1 to occur conditioned by E_2 is denoted with $P(E_1|E_2)$ and has the following formula:

$$P(E_1|E_2) = \frac{P(E_1 \cap E_2)}{P(E_2)}$$

Put more simply, this can be seen as how likely E_1 will occur if E_2 has already occurred.

Regarding conditional probability, the following relationships take place:

- $P(E_1 \cap E_2) = P(E_1) \cdot P(E_2)$

- Bayes theorem

$$P(E_1|E_2) = \frac{P(E_1) \cdot P(E_2|E_1)}{P(E_2)}, P(E_2) > 0$$

An important remark here is that $P(E_1|E_2)$ is different from $P(E_2|E_1)$. In the first relation, E_1 is conditioned by E_2, while in the second relation E_2 is conditioned by E_1.

Random Variables

Let S be a sample space with distribution probability P, and let X be a function that maps S to $\mathbb{R}$. For each element $s_i \in S$, a value $x_i = X(s_i)$ will exist. In this setting, X is called a *random variable*. A random variable gives two important notions in statistics: mean (or expected value) and variance. These values can be computed using these formulas:

- *Mean*: $E(X) = \sum_{s_i \in S} X(s_i) P(s_i) = \sum_i x_i p_i$ with $x_i = X(s_i)$, $p_i = P(s_i)$. Usually, the mean is denoted with μ.

- *Variance*: $Var(X) = E\left((X - \mu)^2\right) = \sum_i (x_i - \mu)^2 p_i$.

Entropy

Let X be a random variable with corresponding values $x_1, ..., x_m$, and $p_i = P(X = x_i)$ for which the following equality occurs:

$$\sum_{i=1}^{n} p_i = 1$$

Let A be an event. In these settings, the entropy of the event A, which gives the uncertainty of A to occur, is computed using this formula:

$$H(A) = -\sum_{j=1}^{m} p_j \lg(p_j) = \sum_{j=1}^{m} p_j \lg\left(\frac{1}{p_j}\right)$$

When $p_i = 0$, the convention is as follows:

$$p_i \lg(p_i) = p_i \lg\left(\frac{1}{p_i}\right) = 0$$

When A, B are two events, the following relations take place:

1. Joint entropy

$$H(A,B) = \sum_{a,b} P(A=a,B=b) \lg(P(A=a,B=b))$$

2. Conditional entropy

$$H(A|B=v) = -\sum_{m} P(A=m|B=v) \lg(P(A=m,B=b))$$

3. Equivocation of A related to B

$$H(A|B) = \sum_{m} P(B=m) H(A|B=m)$$

In the previous relations, a, b go through all of the values within the random variables, A, B respectively, and m goes through all of the values within the random variable A.

A Little Algebra

Prime factorization. This is the process of decomposing a composite number into factors as prime numbers. The prime factorization is unique for each number, according to the fundamental theorem of arithmetic. Prime factorization is important in cryptography, because although it is easy (*easy* meaning a polynomial-time algorithm exists that does this) to check whether a number is prime, there is no general method through which the prime factorization can be achieved. For example, the prime factorization of 4320 is $4320 = 2^5 \bullet 3^3 \bullet 5^1$. Note here that 2, 3, and 5 are prime numbers. Prime factorization is one of the most important hardness assumptions in cryptography. One of the widely used cryptosystems, RSA, lies its hardness assumption on prime factorization.

BigInteger. Big integers are not quite mathematical tools. They are used in computer science (cryptography) to show that an integer value should have a very large value that exceeds the representation of an int. For example, in Java, there is a special class called BigInteger. Listing 6-1 is an example of using the BigInteger class in Java. Figure 6-1 shows the result.

Listing 6-1. Using BigIntegers

```
1    import java.math.BigInteger;
2
3    public class BigIntegerExamples {
4
5        public static void main(String[] args) {
6            BigInteger bi1, bi2;
7
8            //initialization with a string value that will be
                converted to BigInteger
9            bi1 = new BigInteger("5697726366552");
10           System.out.println(bi1);
11
12           //initialization with an int value
13           bi2 = BigInteger.valueOf(87362);
14           System.out.println(bi2);
15
```

```
16              //the operations are implemented as methods within the class
17              BigInteger sum = bi1.add(bi2);
18              System.out.println("Sum: " + sum);
19              BigInteger pow = bi1.pow(bi2.shortValue());
20              System.out.println("Power: " + pow);
21
22              //it provides complex operations such as modulo then power
23              BigInteger result = bi1.modPow(bi2, new BigInteger("5"));
24              System.out.println("Mod inverse: " + result);
25
26              //comparison between two BigIntegers
27              if(bi1.compareTo(bi2) < 0)
28                      System.out.println(bi1 + " less than " + bi2);
29              else
30                      System.out.println(bi1 + " greater than " + bi2);
31
32          }
33      }
```

Figure 6-1. *The result of using BigIntegers*

The RSA problem. From the RSA encryption system, another hardness assumption resulted. The RSA problem is as follows: let (N, e) be the public key of the RSA cryptosystem and C be encrypted text (which is publicly available). It is required to compute efficiently P, where $C \equiv P^e (mod\ N)$. For more about the RSA cryptosystem, see Chapter 10.

Blum integer. A Blum integer is a number of form $n = p \cdot q$, where p and q are prime numbers and they are congruent to 3 mod 4. Listing 6-2 provides a way in which an integer value can be tested as a Blum integer or not. Figure 6-2 shows the result.

Listing 6-2. Testing Whether a Number Is a Blum Integer

```
1    import java.util.Scanner;
2
3    public class BlumInteger {
4        public static boolean checkBlumInteger(int value)
5        {
6            //define an array with elements from 0 to value-1
7            //suppose each number in this interval is prime
8            int primesArray[] = new int[value + 1];
9            for (int i = 0; i < value; i++)
10               primesArray[i] = 1;
11
```

```
12          //update the array with the primality of number within in the
            interval
13          for (int i = 2; i * i <= value; i++) {
14              if (primesArray[i] == 1) {
15                  //if a value is prime, then certainly its multiples
                        are no primes
16                  //update the primality accordingly
17                  for (int j = i * 2; j <= value; j += i)
18                      primesArray[j] = 0;
19              }
20          }

21
22          //check whether the value is a product of two prime numbers
23          //on the same time check whether the value has a form of 4x+3
24          for (int aux = 2; aux <= value; aux++) {
25              if (primesArray[aux] == 1) {
26                  if ((value % aux == 0) && ((aux - 3) % 4) == 0) {
27                      int x = value / aux;
28                      return (x != aux && primesArray[x] == 1  && (x - 3)
                            % 4 == 0);
29                  }
30              }
31          }
32          return false;
33      }

34
35  public static void main(String[] args)
36  {
37      int vals[] = {177, 125};
38      System.out.println("Checking Blum integer...");
39      for(int i=0;i<vals.length;i++)
40      {
41      if (checkBlumInteger(vals[i]) == true)
42          System.out.println(vals[i] + " - True");
43      else
```

```
44                    System.out.println(vals[i] + " - False");
45              }
46         }
47    }
```

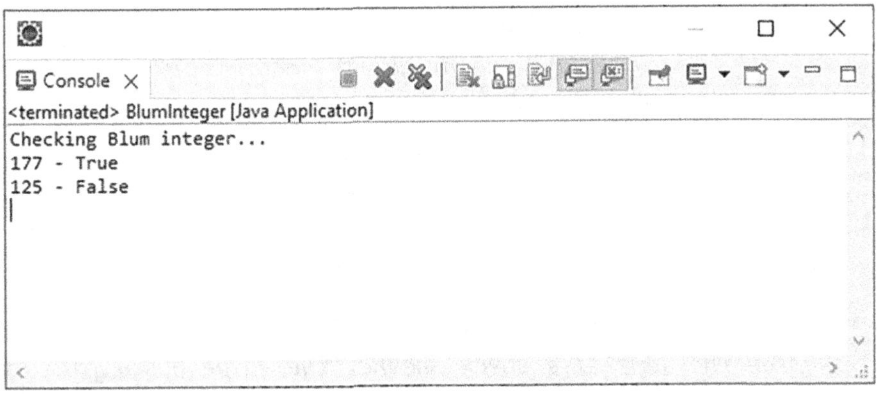

Figure 6-2. *The result of Listing 6-2*

Quadratic residue modulo *n*. This is pretty self-explanatory: it is an integer value *q* for which an integer value *x* exists, such that $x^2 \equiv q (mod\ n)$.

Algebraic structures. These are characterized by the following elements: *A*, which is a set having at least one element and a set of operations defined over the set *A*. The operations must meet some conditions. Examples of algebraic structures are groups, rings, and fields.

Groups. One of the most important elements in cryptography is the group. These are algebraic structures that meet some conditions. Let *G* be a set with at least one element and a binary operation denoted with + defined over the set G. In general, the groups we are working with are abelian. Therefore, the properties that a group has are the following:

- + is well-defined over *G*. This means that $g_1 + g_2 \in G, \forall\, g_1, g_2 \in G$. However, usually, this property is satisfied by default, from the fact that it is a binary operation.

- + is associative. This means that $(g_1 + g_2) + g_3 = g_1 + (g_2 + g_3), \forall\, g_1, g_2, g_3 \in G$.

- Existence of an identity element (unique) $e \in G$ for which
  $g + e = e + g = g, \forall g \in G$.

- Existence of an inverse element (unique) $g' \in G$ for which
  $g + g' = g' + g = e$.

- If $+$ is commutative, then the group is called abelian. $+$ is
  commutative if $g_1 + g_2 = g_2 + g_1, \forall g_1, g_2 \in G$.

Another important algebraic structure used in cryptography is the *field*. A field is a tuple $(K, +, \bullet)$ for which $(1)(K, +)$ is an abelian group, $(2)(K, \bullet)$ is an abelian group, and (3) " $\bullet$ " is distributive over $+$, namely, $k_1 \bullet (k_2 + k_3) = k_1 \bullet k_2 + k_1 \bullet k_3, \quad \forall k_1, k_2, k_3 \in K$.

Let $x_1, ..., x_n \in G$, where G is a group. If for each $x \in G$ some integer values $\alpha_1, ..., \alpha_n$ exist such that

$$x = \alpha_1 x_1 + \alpha_2 x_2 + ... + \alpha_n x_n, \forall x \in G$$

then it is said that G is generated by $S = \{x_1, ...x_n\}$. A group that is generated by only one element is called a *cyclic group*. The *order of a group* is given by the number of its elements. The order of an element $g \in G$ represents the smallest integer m for which $g^m = e$, where $(G, \bullet)$ is the group.

Galois fields. Their name comes from the mathematician Evariste Galois and represents fields with a finite number of elements in the base set. Galois fields are important in computer science because here the data is represented as bits that represent the elements of the Galois field $GF = (Z_2 = \{0, 1\}, +, \bullet)$. Using field properties, different operations can be applied between bits; therefore, data can be scrambled easily. In the general case, a Galois field is defined as follows:

$$GF(p^n) = \{0, 1, ...p-1\} \cup \{p, p+1, ...p+p-1\} \cup \{p^2, p^2+1, ..., p^2+p-1\} \cup ...$$
$$\cup \{p^{n-1}, ..., p^{n-1}+p-1\}$$

In this form, p is a prime number, and n is a positive integer. The values p^n and p are special, as the first gives the order of the field, and the other gives the characteristic of the field. Note that addition and multiplication are made modulo p in a Galois field. The following is an example of GF:

$$GF(2^3) = \{0, 1, 2, 2+1, 2^2, 2^2+1, 2^2+2, 2^2+2+1\} = \{0, 1, 2, 3, 4, 5, 6, 7\}$$

An example of usage Galois fields in computer science is the ASCII code that contains 255 characters. The index of each character can be seen as an element of $GF(2^8)$.

Elliptic Curves

Elliptic curves have applications in different branches of cryptography, especially in digital signatures. These were introduced by Neal Koblitz [1] and Victor Miller [2] independently. Using elliptic curves brings advantages in the terms of efficiency because the size of the keys in an elliptic curves–based cryptosystem is less than other cryptosystems with the same level of security. Elliptic-curve cryptography (ECC) is a branch of public-key cryptography (PKC) based on the algebraic properties of the elliptic curves.

Weierstrauss form for elliptic curves. The graph that corresponds to the following equation is called an elliptic curve E:

$$y^2 + a_1 xy + a_3 y = x^3 + a_2 x^2 + a_4 x + a_6,$$

where a_1, a_2, a_3, a_4, a_6 are constants and $\Delta \neq 0$ (called the discriminant of E). The discriminant has the following form:

$$\Delta = -d_2^2 d_8 - 8 d_4^3 - 27 d_6^2 + 9 d_2 d_4 d_6 \text{ with}$$

$$d_2 = a_1^2 + 4 a_2$$

$$d_4 = 2 a_4 + a_1 a_3$$

$$d_6 = a_3^2 + 4 a_6$$

$$d_8 = a_1^2 a_6 + 4 a_2 a_6 - a_1 a_3 a_4 + a_2 a_3^2 - a_4$$

The condition for delta to be different from 0 is required to eliminate the possibility of multiple roots. When defining an elliptic curve, it should be mentioned from which set the values a_1, a_2, a_3, a_4, $a_6 x$, y are. In cryptography, usually these belong to Z_p (where p

is a prime number) for which there are addition and multiplication operations, such that together they form a field structure. Another possible set for the values is Z_q, where $q = p^k$, $k \geq 1$, and p is prime. When the curve E has the values in a field K, then it is said that E is *defined over K*, and it is denoted with $E(K)$. However, the graph must contain a special point called the *point at the infinite* and noted with ∞, which represents the neutral element. Therefore, $E(K)$ has the following form:

$$E(K) = \{\infty\} \cup \{(x, y) \in K \times K | y^2 + a_1 xy + a_3 y = x^3 + a_2 x^2 + a_4 x + a_6\}$$

For this structure, there is an operation called *point addition*, for which the neutral element is ∞, together forming a group structure $(E(K), +)$. Elliptic-curve cryptography (ECC) lies its hardness assumption on the *discrete logarithm problem (DLP) for elliptic curves*, which states the following:

Given $G = (E(Z_p), +)$, $P \in E(Z_p)$, $Q \in H = \{sP, s \geq 0\}$, compute the unique integer value k, such that $Q = kP$, where $kP = \underbrace{P + \ldots + P}_{k\ times}$.

Note that the operation kP represents the multiplication of a point on the curve with a scalar value. You can find more about elliptic curves arithmetic and its use in cryptography in [3]. Note that DLP can be generalized to any group structure.

One of the most efficient attacks against elliptic curve cryptosystems is the *Pollard-Rho attack* that speculates on the discrete logarithm problem for elliptic curves.

First, it is given the general case of the DLP: let p be a prime number, a a generator for the Diffie-Hellmann problem, x the private key, and y the corresponding public key. Between all of these, the following equality exists:

$$y \equiv a^x \bmod p$$

The general case of DLP requires us to compute x given y, p, a.

The Pollard-Rho attack is a statistical algorithm in which the bit security is divided by two; therefore, it is more efficient than brute-force attacks. While brute-force attacks require n trials, the Pollard-Rho attack requires only $\sqrt{n}$ trials. Its purpose is to find the values $u, v, u', v' \in \{0, \ldots, p-2\}$ for which $a^u y^v = a^{u'} y^{v'}$. From this equality and the equality from earlier (DLP), it results in $a^{u+xv} = a^{u'+xv'}$; therefore, $u + xv = u' + xv' \bmod (p-1)$.

A solution for this equation (i.e., finding x) is found when the equation is solved within the modular ring modulus $p - 1$. Here, one of the following things can happen:

- The solution does not exist. In this situation, other values for u, v, u', v' should be found.

- The solution exists, but it is not unique. In this situation, if there is a relatively small number of solutions, then all of them can be tested by brute force.

Conclusion

This chapter presented beginner background information (definitions and mathematical descriptions) for the world of cryptography. You can find detailed references for mathematical concepts that represent the background of cryptography in any of the following references.

References

[1]. Koblitz, N. (1987). Elliptic curve cryptosystems. Mathematics of computation, 48(177), pp. 203–209.

[2]. Miller, V. S. (1985). Use of elliptic curves in cryptography. In Conference on the theory and application of cryptographic techniques (pp. 417–426). Springer, Berlin, Heidelberg.

[3]. Hankerson, D., Menezes, A. J., Vanstone, S. (2006). Guide to elliptic curve cryptography. Springer Science Business Media.

[4]. Gilbert, L. (2014). Elements of modern algebra. Nelson Education.

[5]. Dent, A. W. (2006). Fundamental problems in provable security and cryptography. Philosophical Transactions of the Royal Society A: Mathematical, Physical and Engineering Sciences, 364(1849), pp. 3215–3230.

[6]. Stinson, D. R., Paterson, M. (2018). Cryptography: theory and practice. CRC press.

[7]. Oppliger, R. (2011). Contemporary cryptography. Artech House.

[8]. Buchmann, J. (2004). Introduction to cryptography (Vol. 335). New York: Springer.

[9]. Menezes, A. J., Van Oorschot, P. C., & Vanstone, S. A. (2018). Handbook of applied cryptography. CRC press.

[10]. Etienne, E. (2019). Elementary statistical methods of cryptography.

[11]. Mihailescu, M. I., & Nita, S. L. (2021). Pro Cryptography and Cryptanalysis with C++ 20: Creating and Programming Advanced Algorithms. Apress.

CHAPTER 7

Pseudorandom Number Generators

Randomness is a key element in cryptography, as cryptosystems and protocols are based on arbitrary numbers that cannot be predicted. Numbers with such characteristics are known as randomly generated numbers (or random numbers), and the technique of their generation lies in a statistical context. Random numbers are mainly used in domains in which results that cannot be predicted are expected, for example, statistical sampling, simulations, or even gambling.

There are two main ways in which random numbers can be generated. The first way is based on phenomenons from physics that prove to be random, generating great entropy values. The second way is to create algorithms that can "choose"/ compute a sequence of numbers that appears to be random. Such algorithms are called *pseudorandom numbers generators* (PRNGs), and usually, their base is a special value called a *seed* from which the random generation of numbers starts in a specific way. Because the PRNGs use this initial value seed, they are not truly random, which is the reason for the *pseudo* in their name. However, the seed may come from natural sources and may be changed occasionally. Another way to generate pseudorandom numbers consists of combining the two main approaches.

Statistically, there is not too much information about what "random bits" refers to, but from the computing machine point of view, *random bits* refers to a sequence of bits chosen randomly.

To be considered a PRNG, an algorithm should meet the following conditions:

- It is characterized by simplicity and fastness.

- The sequence of numbers that it generates should have an arbitrary length, and the numbers should not repeat. Limited by the technology, a computer (or another device) can generate numbers

© Stefania Loredana Nita and Marius Iulian Mihailescu 2022
S. L. Nita and M. I. Mihailescu, *Cryptography and Cryptanalysis in Java*,
https://doi.org/10.1007/978-1-4842-8105-5_7

only smaller than a value, even if that value is the maximum (highest) number that the computer supports; therefore, the period cannot be infinite. However, the period can be set as high as possible.

- The numbers should not depend on each other.

- The generated numbers should be uniformly distributed from the statistical point of view.

To be used in cryptography, a PRNG should have the following additional properties:

- *Next-bit test* [1]: Fixing an arbitrary integer k, an attacker cannot predict the value of the bit positioned on the $(k + 1)$ index of the sequence using a polynomial-time algorithm if it knows the values of the first k bits but does not know the seed. In [1], the author proved that a PRNG passing the next-bit test will pass all statistical tests for randomness in polynomial time.

- *State compromise extensions*: Breaking a subsequence of the pseudorandom sequence of numbers, an attacker cannot recover the entire sequence.

In statistics, there is still some information about random bits: the frequency of the appearance of 1 and 0 should be the same; the frequency for 00, 11, 01, 10 should be the same, but these should appear half as less often than 1 and 0. To test "how much" a sequence of numbers is random, different statistical tests may be used, such as Diehard [2] or Kolmogorov-Smirnov [3]. Randomness is an essential aspect used in cryptography, and deducing or recovering a random number should be very difficult (ideally impossible). To be considered a perfect random number, the number may be recovered by an attacker only by using brute force. Many techniques from cryptanalysis exploit the weaknesses of the functions that generate random numbers.

Depending on the purpose, the required randomness degree of the PRNG can be different. For example, when keys of an encryption system are generated, the degree should be very high, while in protocols uniqueness for the generated nonce is required and the degree of randomness is not required to be very high.

Some standards that describe the PRNGs are FIPS 186-4 [4], NIST SP 800-90A [5], ANSI X9.17-1985 Appendix C, ANSI X9.31-1998 Appendix A.2.4, and ANSI X9.62-1998 Annex A.4, which is obsoleted by ANSI X9.62-2005, Annex D (HMAC_DRBG) [6].

Examples of PRNGs

In this section, we present some PRNGs that are not necessarily as important as PRNGs themselves but are important from the technical point of view and the mechanisms used, because they represent the basis for other more complex or powerful PRNGs.

Linear Congruential PRNGs

In [7], the author introduced a simple linear congruential generator, given by the following recurrence:

$$r_{n+1} = ar_n + b\,(mod\ m),$$

with a, b, $m \in \mathbb{N}$ constants and r_0 being the seed. A general formula for linear congruential PRNGs is as follows:

$$r_{n+k} = \left(a^k r_n + \left(a^k - 1\right)cb^{-1}\right)(mod\ m)$$

In these types of PRNGs, cycles occur, with the maximum length (called a *period*) of cycles being m. The advantage of linear congruential PRNGs is the fastness of computations. Its general form achieves the maximum period in the following situations:

- $\gcd(c, m) = 1$
- $b = a - 1$ is multiple of a prime number that divides m
- $b\ mod\ 4 = 0$ if $m\ mod\ 4 = 0$

Linear congruential PRNGs cannot be used in cryptography because they are very vulnerable: if an attacker can discover the seed, then it breaks the generator. Comprehensive cryptanalysis is presented in [8].

Blum-Blum-Shub PRNG

A simple and efficient PRNG is Blum-Blum-Shub (BBS), also called a *residual generator* [9], based on Blum integers (see Chapter 6 for a description of a Blum integer). To generate random numbers using BBS, one should proceed as follows [9]:

- Establish an arbitrary integer k, choose two prime numbers p, q represented on $k/2$ bits, and let $n = pq$.

- Compute x_0 as a quadratic residue modulo n and define the series $r_{i+1} = r_i^2 \pmod{n}$.

- Set $s_i = r_i \pmod 2$, $1 \leq i \leq m$ and generate the number $f(x_0) = s_1 s_2 \ldots s_m$, where m is the length of the series at an arbitrary moment.

As s_i can be determined independently as $s_i = x_0^{2^i (mod\,(p-1)(q-1))} \bmod 2$, the BBS PRNG does not lay into recursion. BBS is constructed on the hard problem of factoring integers, in this case, factoring n. Integer n can be public, as there is no restriction for who may generate pseudorandom numbers, but n should be difficult to be factored (ideally impossible), in which case the value that is generated cannot be predicted. Another important feature of BBS is that if an attacker obtains a subsequence of numbers generated by BBS PRNG, then it cannot predict the previous bit or the next bit. Although it is a good PRNG, BBS is inefficient. A more efficient variation of it is presented in [10], in which considering l the length of x_i, then the last $\lfloor \log_2 x_i \rfloor$ are used further. In cryptography, BBS is one of the best PRNGs used in protocols for generating and distributing the keys involved in a cryptosystem.

Linear Circuit PRNGs

Error detection and correction theory and linear cryptosystems (for example, AES) use linear circuits (also called *shift registers*) because the computations are fast. In short, the linear circuit, called a *linear feedback shift register* (LFSR), is a serial register with a feedback function. An n–LFSR is a register with n flip-flops of data (DF-F). You can find more information about shift registers in [11].

A Geffe generator [12] is a PRNG that uses three LFSRs to combine their nonlinearity. However, a Geffe generator is not resistant against a correlation attack, because the number generated by the generator is the same as the output of the second LFSR in 75 percent of cases. Knowing the polynomial representation of the LFRSs, it would be easy to compute the seed and the series produced by the second LFRS and then to compute the frequency of when the result produced by the second LFSR is the same as the result produced by the PRNG. A similar deduction can be made for the third LFSR, for which the Geffe PRNG also produces similar results in 75 percent of cases. Correlating these two pieces of information, then the entire sequence of generated numbers can be recovered [12].

Other types of LFSR generators are called *stop-and-go*. Some examples of stop-and-go PRNGs are the Beth-Piper PRNG [13] (based on three circuits for which the clocks are controlled; however, it does not resist against correlation attack) and the Gollmann PRNG [14] (based on a serial construction of LFSRs, in which the clock of the current LFRS is controlled by the previous LFSR).

Other PRNGs

The PRNGs lie their difficulty of recovering the sequence or predicting the next value in different mathematical hardness assumptions.

An example of such a mathematical problem is the discrete logarithm problem that represents the base for the Blum-Micali PRNG [15]. The setup for it consists of p, g, which are prime numbers, and an initial value of x_0. A sequence of random numbers is given by the formula $x_{i+1} = g^{x_i} \pmod{p}$. If $x_i < \dfrac{p-1}{2}$, then the exit is 1; otherwise, the exit is 0. For a large prime number p, the Blum-Micali PRNG is secure.

Another example of a mathematical problem that can be used in PRNGs is the RSA problem [16], which represents the hardness of breaking the RSA cryptosystem. The initial setup for this PRNG consists of the following values: p, q for large prime numbers, $n = pq$; the value e that accomplishes the following criteria $gcd(e, (p-1)(q-1)) = 1$; and an initial value $x_0 < n$. A sequence of random numbers is given by the formula $x_{i+1} = x_i^e \pmod{n}$. The exit of the RSA PRNG is $z_i = x_1(mod\, 2)$. For a large number n, the generator is secure.

Another type of PRNG is the Ranrot class [17] used in Monte Carlo algorithms, which use Fibonacci numbers in their generation along with a shifting operation. Mainly, there are four types of Ranrot generators.

PRNGs Security

As you have already seen, pseudorandom number generators are extremely important in cryptography. Because these are used in many cryptographic techniques, the PRNGs themselves should be secure. [18] is a good reference for attacks applied to PRNGS, as the author explores a wide range of attacks, grouping them into these three categories:

- *Direct cryptanalytic attacks*: In these types of attacks, the attacker can distinguish between the output values of the PRGN and the random outputs. This category of attacks can be applied to almost all use cases of PRNGs.

- *Input-based attacks*: In these types of attacks, the attacker can use information about the input values or control them in order to find the vulnerabilities of the PRNG. This also leads to distinguishing between the output values of the PRGN and the random outputs.

- *State compromise extension attacks*: In these types of attacks, the attacker can use further information about the internal state of the PRNG. These attacks can be launched in situations in which at a certain time the internal state of the PRNG was compromised (even for a short time) and the attacker is able to collect further information or take further advantages, although the situation was fixed.

A detailed description of these classes of attacks can be found in [18], which also presents examples of attacks that can be applied to known PRNGs. A more recent work about PRNGs security is [19].

Java.util.Random Class

In Java, random numbers can be generated using the `java.util.Random` class [20], which can be instantiated using one of the following constructors:

- `Random()`: The seed for this generator is a value that was not used in any other calls to instantiate in this way.

- `Random(long seed)`: The seed is given by the user.

Using this class, you can generate streams of `int, double,` and `long` random numbers or `int, double, float, long,` and `boolean` values. The seed is represented in 48 bits but then changed through a linear congruential relationship. Note that the individual values are generated using a uniform distribution, which means any value has (almost) the same chances to be generated as the rest of the values of the same type. See Listing 7-1.

Listing 7-1. Java.util.Random

```
1    import java.util.Random;
2
3    public class RandomExamples {
4
5            public static void main(String[] args) {
6
7            Random randomObject = new Random();
8            System.out.println("Generating an integer value < 1000... " +
                 randomObject.nextInt(1000));
9            System.out.println("Generating an integer value... " +
                 randomObject.nextInt());
10           System.out.println("Generating a long value... " +
                 randomObject.nextLong());
11           System.out.println("Generating a Boolean value... " +
                 randomObject.nextBoolean());
12           System.out.println("Generating a double value... " +
                 randomObject.nextDouble());
13           System.out.println("Generating a float value... " +
                 randomObject.nextFloat());
14           System.out.println("Generating an integer number... " +
                 randomObject.nextGaussian());
15
16           byte[] byteArray = new byte[15];
17           randomObject.nextBytes(byteArray);
18           System.out.println("Generating a sequence of bytes...");
19           System.out.print("[");
20           for(int i = 0; i< byteArray.length; i++)
21           {
22               System.out.print(byteArray[i] + " ");
23           }
24           System.out.print("]\n");
25       }
26    }
```

The code from Listing 7-1 is straightforward, and it generates different types of random values. Figure 7-1 shows the output.

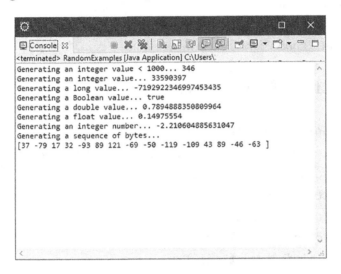

Figure 7-1. *Output for Listing 7-1*

However, the instances of Random are not cryptographically secure. To use secure PRNGs, Java provides a special class called java.security.SecureRandom [21]. The PRNGs in this class are implemented according to FIPS 140-2 and RFC 1740 standards. The ways in which the objects of the SecureRandom class are used are similar to how Random is used. In Listing 7-2 we adapted the previous example to generate securely random values. Figure 7-2 shows the output.

Listing 7-2. Generating Securely Different Types of Values

```
1   import java.security.SecureRandom;
2
3   public class SecureRandomExample {
4
5           public static void main(String[] args) {
6           SecureRandom secureRandomObject = new SecureRandom();
7
8           System.out.println("Securely generating an integer value <
            1000... " + secureRandomObject.nextInt(1000));
```

```
9        System.out.println("Securely generating an integer value... " +
         secureRandomObject.nextInt());
10       System.out.println("Securely generating a long value... " +
         secureRandomObject.nextLong());
11       System.out.println("Securely generating a Boolean value... " +
         secureRandomObject.nextBoolean());
12       System.out.println("Securely generating a double value... " +
         secureRandomObject.nextDouble());
13       System.out.println("Securely generating a float value... " +
         secureRandomObject.nextFloat());
14       System.out.println("Securely generating an integer number... " +
         secureRandomObject.nextGaussian());
15
16       byte[] byteArray = new byte[15];
17       secureRandomObject.nextBytes(byteArray);
18       System.out.println("Securely generating a sequence of
         bytes...");
19       System.out.print("[");
20       for(int i = 0; i< byteArray.length; i++)
21       {
22           System.out.print(byteArray[i] + " ");
23       }
24       System.out.print("]\n");
25   }
26 }
```

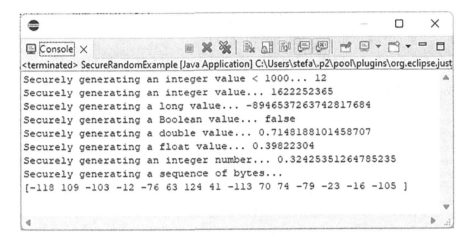

Figure 7-2. *Output for Listing 7-2*

The fundamental difference between java.util.Random and java.security. SecureRandom is given by how the seed is chosen. In java.util.Random the seed is generated based on the clock of the system, while in java.security.SecureRandom the seed is generated based on random data from the operating system.

Further, several secure coding practices are recommended when working with PRNGs.

- Use PRNGs already implemented, tested, widely used, and standardized. In this way, most of the possible vulnerabilities are fixed, and a high degree of security is ensured.

- Use PRNGs with strong security in applications; use, for example, java.security.SecureRandom instead of java.util.Random. However, the goal for which PRNG is used is very important. There are situations in which the applications need faster PRNGs rather than strong, secure PRNGs. On the other hand, other applications will require strong secure PRNGs rather than efficiency.

- Use a set of generated values only one time. Each time a series of pseudorandom numbers is needed, the best approach is to generate a new sequence.

- The values generated by the PRNG should be secured. If they should be stored somewhere, in a database or on a server, make sure that the database or the file that contains the generated values are strongly secured.

Conclusion

In this chapter, we discussed pseudorandom number generators. You learned which characteristics an algorithm needs to have to be considered a pseudorandom generator. In addition, to be used in cryptography, the pseudorandom generator must have two additional characteristics. Further, we presented some PRNGs and the technical aspects behind them, and lastly, we provided straightforward examples of how `java.util.Random` and `java.security.SecureRandom` can be used.

References

[1]. Yao, A. C. (1982, November). Theory and application of trapdoor functions. In 23rd Annual Symposium on Foundations of Computer Science (SFCS 1982) (pp. 80–91). IEEE.

[2]. Marsaglia, G. (1996). DIEHARD: a battery of tests of randomness. Available online: `http://stat.fsu.edu/geo`.

[3]. Massey Jr, F. J. (1951). The Kolmogorov-Smirnov test for goodness of fit. Journal of the American Statistical Association, 46(253), 68–78.

[4]. FIPS PUB 186-4, Available online: `https://nvlpubs.nist.gov/nistpubs/FIPS/NIST.FIPS.186-4.pdf`

[5]. NIST 800-90A Rev. 1, Available online: `https://csrc.nist.gov/publications/detail/sp/800-90a/rev-1/final`

[6]. Random Bit Generation, Available online: `https://csrc.nist.gov/projects/random-bit-generation`

[7]. Lehmer, D. H. (1951). Mathematical methods in large-scale computing units. Annu. Comput. Lab. Harvard Univ., 26, 141–146.

[8]. Plumstead, J. B. (1982, November). Inferring a sequence generated by a linear congruence. In 23rd Annual Symposium on Foundations of Computer Science (sfcs 1982) (pp. 153–159). IEEE.

[9].   Blum, L., Blum, M., & Shub, M. (1986). A simple unpredictable pseudo-random number generator. SIAM Journal on computing, 15(2), 364–383.

[10].   Sidorenko, A., & Schoenmakers, B. (2005, December). Concrete security of the Blum-Blum-Shub pseudorandom generator. In IMA International Conference on Cryptography and Coding (pp. 355-375). Springer, Berlin, Heidelberg.

[11].   Goresky, M., & Klapper, A. (2012). Algebraic shift register sequences. Cambridge University Press.

[12].   Wei, S. (2006). On generalization of Geffe's generator. IJCSNS International Journal of Computer Science and Network Security, 6(8A), 161–165.

[13].   Beth, T., & Piper, F. (1984, April). The Stop-and-Go Generator. In Eurocrypt (Vol. 84, pp. 88–92).

[14].   Park, S. J., Lee, S. J., & Goh, S. C. (1995, August). On the security of the Gollmann cascades. In Annual International Cryptology Conference (pp. 148-156). Springer, Berlin, Heidelberg.

[15].   Blum, M., & Micali, S. (1984). How to generate cryptographically strong sequences of pseudorandom bits. SIAM journal on Computing, 13(4), 850–864.

[16].   Friedlander, J. B., Lieman, D., & Shparlinski, I. E. (1999). On the distribution of the RSA generator. In Sequences and their applications (pp. 205–212). Springer, London.

[17].   Fog, A. (2001). Chaotic random number generators with random cycle lengths. Available online: http://www.agner.org/random/theory/chaosran.pdf

[18].   Kelsey, J., Schneier, B., Wagner, D., & Hall, C. (1998, March). Cryptanalytic attacks on pseudorandom number generators. In International workshop on fast software encryption (pp. 168–188). Springer, Berlin, Heidelberg.

[19]. Ruhault, S. (2017). SoK: Security models for pseudo-random number generators. IACR Transactions on Symmetric Cryptology, 506–544.

[20]. Class Random, `https://docs.oracle.com/javase/8/docs/apiww/java/util/Random.html`

[21]. Class SecureRandom, `https://docs.oracle.com/javase/8/docs/api/java/security/SecureRandom.html`

CHAPTER 8

Hash Functions

In information security, hash functions are an important cryptographic tool. They have applications in a wide range of use cases, such as securing passwords in a database, securing authentication, and maintaining message integrity.

In a few words, a *hash function* is a compression mathematical function that takes an input with any length and converts it into a string with a fixed length. The input can be almost anything such as text or different types of files. The output, on the other hand, should be unique for every input and has a predefined fixed length according to the type of the hash function. The output is called a *message digest, hash value,* or *hash.*

A hashing class is a tuple with four elements, as follows [1]: the set of all possible messages (encrypted or not) X; a finite set of fingerprints (digests) Y; a finite set of keys K; and a set H of hash functions such that for every $k \in K$ there exists a hash function $h_K \in H$, $h_K : X \rightarrow Y$. A pair $(x, y) \in X \times Y$ is valid for the key K if $h_K(x) = y$.

The following problems should be hard to compute [1]:

- *Noninvertible*: For a given $y \in Y$, it is hard to discover $x \in X$, such that $y = h(x)$. This property means that it is hard to invert the function h, and it ensures that if an attacker has an output value, then it cannot compute the input value.

- *Weak collisions*: For a given pair (x, y), it is hard to discover $x_1 \neq x$ such that $h(x_1) = h(x)$. This property protects against situations in which the attacker has the input value and the corresponding hash and tries to modify legitimate values with other values that seem to be legitimate.

- *Collisions*: It is hard to find two different values $x, x_1 \in X$ such that $h(x) = h(x_1)$.

By "hard" we mean that solving the problem involves only algorithms of high complexity (eventually, the problem is NP-complete). A hash function $h : X \rightarrow Y$ for

© Stefania Loredana Nita and Marius Iulian Mihailescu 2022
S. L. Nita and M. I. Mihailescu, *Cryptography and Cryptanalysis in Java,*
https://doi.org/10.1007/978-1-4842-8105-5_8

which the collision problem is difficult is called a *collision-resistant* hash function. Because $\#X > \#Y$, any hash function admits collisions, but the trick is to make it extremely difficult to find the collisions and only by using algorithms with high complexity. (We denoted with $\#A$ the cardinal of the set A, namely, the number of elements contained by the set.)

It is enough that a hash function is collision resistant because this property implies weak collision and noninvertibility.

Figure 8-1 shows an example of a hash function. The plain message is "Sample Plaintext," and after applying the hash function SHA-512 to it, the text on the right is obtained.

Figure 8-1. *Applying SHA-512 on text*

Several theorems establish an inferior limit for the cardinal of Y. One of them is the birthday paradox through which a relation between $\#X$ and $\#Y$ is determined. The relation is as follows: $\#X \geq 2 \bullet \#Y$. The birthday attack is a brute-force attack that follows to discover collisions. Generally, an attacker randomly generates some messages x_1, x_2, ... $\in X$ (in practice, they use pseudorandom number generators) and, for each x_i computes and stores the digest $y_i = h(x_i)$. Then, the attacker compares y_i with the ones stored previously. If y_i equals another stored y_j, then the attacker found a collision whose form is (x_i, x_j). According to the birthday paradox, this would happen after about $2^{\frac{N}{2}}$ messages. A hash function that represents the hash value on 40 bits is vulnerable to the birthday attack using only 2^{20} (approximately one million) random messages. Therefore, in practice, it is recommended to use hash functions that represent the values on at least 256 bits. Table 8-1 presents a list of keyed cryptographic hash functions, while Table 8-2 presents a list of unkeyed cryptographic functions [2].

Table 8-1. *Keyed Cryptographic Functions*

Name	Length of the Tag (Bits)	Type	References
BLAKE2	Arbitrary	Keyed hash function with prefix-MAC	[3], [4]
BLAKE3	Arbitrary	Keyed hash function with supplied initializing vector (IV)	[5]
HMAC	-	-	[6]
KMAC	Arbitrary	Based on Keccak	[7], [8]
MD6	512	Merkle tree with NLFSR	[9]
PMAC	-	-	[10]
UMAC	-	-	[11]

Table 8-2. *Unkeyed Cryptographic Functions*

Name	Length of the Tag (Bits)	Type	References
BLAKE256	256	HAIFA structure [12]	[15]
BLAKE512	512	HAIFA structure [12]	[15]
gost	256	Hash	[16]
md2	128	Hash	
md4	128	Hash	[17]
md5	128	Merkle-Damgard construction [13]	[18]
md6	Up to 512	Merkle tree NLFSR	[9]
ripemd	128	Hash	[19], [21]
ripemd-128	128	Hash	[19]
ripemd-160	160	Hash	[20]
ripemd-256	-	Hash	[21]
ripemd-320	320	Hash	[22]
sha-1	160	Merkle-Damgard construction [13]	[23]
sha-224	224	Merkle-Damgard construction [13]	[25]
sha-256	256	Merkle-Damgard construction [13]	[24]
sha-384	384		[24]
sha-512	512		[24]
sha-3 (keccak)	Arbitrary	Sponge functions [14]	[26], [27]
whirlpool	512	Hash	[28], [29]

Hash functions have good applicability in digital signatures. Usually, when working with digital signatures, the hash value of a long message is computed; only the hash value is signed, and the sender sends it to the receiver. On the other hand, the receiver computes the hash value of the received message. If it corresponds to the one sent by the sender, then the message is righteous. This approach saves time and space complexity because, otherwise, the original message should be split into blocks of a predefined length, and then each block needs to be signed.

The BLAKE family [15] of hash functions uses the ChaCha stream cipher [30] as its mathematical basis. The modifications besides ChaCha consist of adding a permuted version of the input block XORed with constant numbers of the current round before each ChaCha round. In BLAKE, there are used hash values of 8 words, which are combined with 16 message words, and then the ChaCha value is truncated, obtaining the next hash value.

The GOST hash function works as follows: the input message is split into blocks of 256 bits, and then a number of 0s are padded to the message such that it reaches the length of 256 bits. The bits that are remaining are replaced by the sum of all previous hashed blocks represented on 256 bits and an integer value represented on 256 bits that is the length of the original message. However, [31] describes a collision attack that breaks it into 2^{105} time.

The father of the message digest (MD) hash functions is Ronald Rivest, who proposed MD2 in 1989. MD2-MD5 hash functions work on 128 bits. Each represents an improvement of the previous one, and their mathematical construction is based on the Merkle-Damgard function. Although on time, it was proven that they are not resistant to collision attacks. For example, MD5 is still used as a checksum for data integrity, although it is vulnerable to length extension attacks [32]. However, MD hash functions represent the basis for other hash functions, such as RIPEMD. MD6 uses another approach, namely, the Merkle tree. Their authors show that MD6 resists different types of differential attacks [33].

The RACE Integrity Primitives Evaluation Message Digest (RIPEMD) family of functions are created based on MD4 and MD5, and their performance is close to the one of SHA-1.

The Secure Hashing Algorithms (SHA) family of hash functions is issued by the National Institute of Standards and Technology (NIST). SHA-1 was designed by the National Security Agency (NSA) after it was proved that the original SHA was insecure. The basis of SHA-1 lies in the MD5 hash function. However, it was proved that SHA-1 is also insecure. SHA-224, SHA-256, SHA-384, and SHA-512 are included in the next iteration of the SHA functions (known as SHA-2), which were also designed by the NSA. SHA-3 hash functions represent the results of authors from outside of the NSA. These are a subset of the Keccak family primitives, whose mathematical basis lies in sponge functions. The reference [34] presents a well-organized table regarding the security of the SHA hash functions. Here it is shown that SHA-512 and SHA-3 are secure against collision attacks and length extension attacks.

The whirlpool hash function outputs a digest of 512 bits, and its mathematical basis lies on the square block cipher, which was also used to design the AES encryption system.

The most popular hash functions are those from the SHA-2 family. Although the MD family and others like it are not secure anymore, they can be widely used as checksum functions.

Although the hash functions are widely used with applications such as securing passwords in databases, checking the integrity of different types of files, signing in digital signatures, etc., they have a more recent important application, namely, the blockchain. The entire infrastructure of blockchain technology lies in hash functions. Blockchain, which is also called a *distributed ledger,* can be seen as a decentralized database; therefore, there is no central authority that manages the database. All nodes within the network are communicating on a peer-to-peer basis, which means that one node communicates with all of the rest directly. One of the elements that a block of data contains is the hash value of the previous block. With different types of consensus algorithms that establish which node can add a block in the blockchain and this approach of the structure, the data within a blockchain structure is extremely difficult to forge. Figure 8-2 shows the general structure of a chain. Observe how the arrows indicate that the field Prev contains the hash value of the previous data block. For example, Bitcoin uses SHA-256 as its hash function.

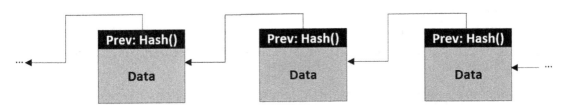

Figure 8-2. *The structure of a chain*

In Java, using a hash function is easy, because they are already implemented in Java libraries (see more details in Chapter 2). In Listing 8-1 we provide a simple example of applying SHA-512 over input text. Figure 8-3 shows the result. Note that even a slight modification of the input text yields a very different hash value. In Figure 8-3 the word *hash* is written with a lowercase *h*, while in Figure 8-4, the word *Hash* is written with an uppercase *H*.

Listing 8-1. Example of Applying SHA-512 on Input Text

```
1   import java.security.MessageDigest;
2   import java.util.Scanner;
3
4   public class HashFunctionExample {
5     public static void main(String args[]) throws Exception
6     {
7        System.out.println("\n Type the text: ");
8
9        Scanner scn = new Scanner(System.in);
10       String input = scn.nextLine();
11       scn.close();
12
13       System.out.println("\n The input text: " + input);
14
15       MessageDigest output_sha = MessageDigest.getInstance("SHA-512");
16
17       output_sha.update(input.getBytes());
18
19
20       byte[] digest = output_sha.digest();
21       System.out.println(digest);
22
23
24       StringBuffer hex_digest = new StringBuffer();
25
26       for (int i = 0;i<digest.length;i++)
27       {
28         hex_digest.append(Integer.toHexString(0xFF & digest[i]));
29       }
30       System.out.println("The hex representation : " + hex_digest.
         toString());
31     }
32   }
```

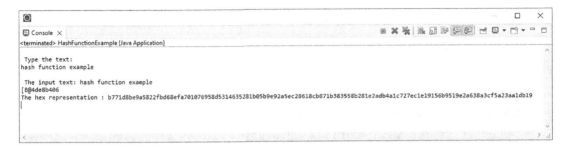

Figure 8-3. *The result of Listing 8-1 for the input "hash function example"*

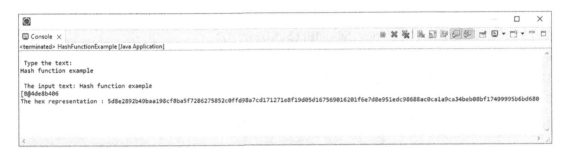

Figure 8-4. *The result of Listing 8-1 for the input "Hash function example"*

The three properties of hash functions (pre-image resistance, second pre-image resistance, collision resistance) give the security level of hash functions. These ensure that if an attacker modifies the input value (even a slight change), then the hash value of the input is also modified. Thus, this shows if the original value has integrity untouched. Different types of collisions can occur [36].

- *Near-collision*: For two input values, only some parts of the corresponding hash values are common.

- *Start-collision*: There can be various types of start conditions that are changed, for example, an initial value.

Besides collision resistance, hash functions must also meet another two security requirements.

- *Indistinguishability*: An attacker cannot distinguish a cryptographic primitive (for example, a block cipher) from an ideal one (for example, a permutation).

- *Indifferentiability*: This type of security is more suitable for hash functions than indistinguishability, due to hash function constructions and purpose. Indifferentiability targets the interaction between cryptographic primitives.

In [35] and [36], the authors classify attacks on hash functions into two categories.

- *Generic attacks*: These attacks target different properties of a hash function, except for collision resistance. Once a vulnerability is found, then the attacker can exploit it. These types of attacks do not depend on the details of the hash function's construction. Examples of such attacks are exhaustive search (an input value is chosen, and then the attacker searches through all possible hash values to find a match), (memoryless) birthday attacks, time-memory trade-off attacks (here, given a value, an attacker can compute the preimage), and meet-in-the-middle (this attack can be launched on an easily invertible hash function).

- *Cryptanalytic attacks*: These types of attacks target the compression function that is the foundation of a hash function. Such attacks have been launched for MD5 and SHA-1. Examples of such attacks are fixed-point attacks (which means that for a given message, if an attacker finds a component of the message, then the message without that component has the same hash value as the original one), length-extension attacks (this attack involves computing the hash function of an extended version of an original message), multicollision attacks (based on the fact that more input values have the same hash value), and herding attacks (the attacker can compute the preimage of an unknown yet hash value).

Conclusion

In this chapter, hash functions were presented. Hash functions are widely used cryptographic primitives, and they have a large number of applications in the field of information security.

A newer use case for hash functions is blockchain technology, which bases its core on hash functions.

References

[1].   Atanasiu, A. (2009). Securitatea Informatiei vol 2 (Protocoale de securitate). INFODATA Cluj Publishing House.

[2].   Mihailescu, M. I., & Nita, S. L. (2021). Pro Cryptography and Cryptanalysis: Creating Advanced Algorithms with C# and. NET. Apress.

[3].   Benaloh, J. (Ed.). (2014). Topics in Cryptology--CT-RSA 2014: The Cryptographer's Track at the RSA Conference 2014, San Francisco, CA, USA, February 25-28, 2014, Proceedings (Vol. 8366). Springer.

[4].   BLAKE2 Official Implementation. Available: https://github.com/BLAKE2/BLAKE2

[5].   Blake3. Available: https://github.com/BLAKE3-team/BLAKE3/

[6].   H. Krawczyk, M. Bellare, R. Canetti, "HMAC: Keyed-Hashing for Message Authentication," RFC 2104, 1997.

[7].   API KMAC. Available: www.cryptosys.net/manapi/api_kmac.html.

[8].   Kelsey, J., Chang, S. J., & Perlner, R. (2016). SHA-3 derived functions: cSHAKE, KMAC, TupleHash, and ParallelHash. NIST special publication, 800, 185.

[9].   Rivest, R. L., Agre, B., Bailey, D. V., Crutchfield, C., Dodis, Y., Fleming, K. E., ... & Yin, Y. L. (2008). The MD6 hash function–a proposal to NIST for SHA-3. Submission to NIST, 2(3), 1-234. Available online: http://groups.csail.mit.edu/cis/md6/submitted-2008-10-27/Supporting_Documentation/md6_report.pdf

[10].   PMAC. Available online: https://web.cs.ucdavis.edu/~rogaway/ocb/pmac.htm

[11].   UMAC. Available online: http://fastcrypto.org/umac/

[12].   Biham, E., & Dunkelman, O. (2007). A Framework for Iterative Hash Functions---HAIFA (No. CS Technion report CS-2007-15). Computer Science Department, Technion.

[13]. Damgård, I. B. (1989, August). A design principle for hash functions. In Conference on the Theory and Application of Cryptology (pp. 416–427). Springer, New York, NY.

[14]. Gilbert, H., & Handschuh, H. (2003, August). Security analysis of SHA-256 and sisters. In International workshop on selected areas in cryptography (pp. 175–193). Springer, Berlin, Heidelberg.

[15]. BLAKE-256. Available online: `https://docs.decred.org/research/blake-256-hash-function/`

[16]. GOST. Available: `https://tools.ietf.org/html/rfc5830`

[17]. Rivest, R. L. (1990, August). The MD4 message digest algorithm. In Conference on the Theory and Application of Cryptography (pp. 303-311). Springer, Berlin, Heidelberg.

[18]. Rivest, R., & Dusse, S. (1992). The MD5 message-digest algorithm.

[19]. RIPEMD-128. Available online: `https://homes.esat.kuleuven.be/~bosselae/ripemd/rmd128.txt`

[20]. RIPEMD-160. Available online: `https://homes.esat.kuleuven.be/~bosselae/ripemd160.html`

[21]. Yeh, Y. S., & Chou, J. S. (2001). Keyed/Unkeyed RIPEMD-128, 192, 256. Journal of Information and Optimization Sciences, 22(3), 563–578.

[22]. Sasaki, Y., & Aoki, K. (2009, July). Meet-in-the-middle preimage attacks on double-branch hash functions: Application to RIPEMD and others. In Australasian Conference on Information Security and Privacy (pp. 214–231). Springer, Berlin, Heidelberg.

[23]. Wang, X., Yin, Y. L., & Yu, H. (2005, August). Finding collisions in the full SHA-1. In Annual international cryptology conference (pp. 17–36). Springer, Berlin, Heidelberg.

[24]. Descriptions of SHA-256, SHA-384, and SHA-512. Available online: `www.iwar.org.uk/comsec/resources/cipher/sha256-384-512.pdf`.

[25]. A 224-bit One-way Hash Function: SHA 224. Available online: https://datatracker.ietf.org/doc/html/rfc3874

[26]. Paul Hernandez, "NIST Releases SHA-3 Cryptographic Hash Standard," 2015.

[27]. Morris J. Dworkin, "SHA-3 Standard: Permutation-Based Hash and Extendable-Output Functions". Federal Inf. Process. STDS. (NIST FIPS) – 202. 2015.

[28]. Paulo S. L. M. Barreto "The WHIRLPOOL Hash Function". 2008. Archived from the original on 2017-11-29. Retrieved 2018-08-09.

[29]. Paulo S. L. M. Barreto and Vincent Rijmen, "The WHIRLPOOL Hashing Function." 2003. Archived from the original (ZIP) on 2017-10-26. Retrieved 2018-08-09.

[30]. Bernstein, D. J. (2008, January). ChaCha, a variant of Salsa20. In Workshop record of SASC (Vol. 8, pp. 3–5).

[31]. Mendel, F., Pramstaller, N., Rechberger, C., Kontak, M., & Szmidt, J. (2008, August). Cryptanalysis of the GOST hash function. In Annual International Cryptology Conference (pp. 162-178). Springer, Berlin, Heidelberg.

[32]. Stevens, M. (2006). Fast Collision Attack on MD5. IACR Cryptol. ePrint Arch., 2006, 104.

[33]. Rivest, R. L., Agre, B., Bailey, D. V., Crutchfield, C., Dodis, Y., Fleming, K. E., ... & Yin, Y. L. (2008). The MD6 hash function–a proposal to NIST for SHA-3. Submission to NIST, 2(3), 1-234.

[34]. Secure Hash Algorithms, Available online: https://en.wikipedia.org/wiki/Secure_Hash_Algorithms

[35]. Gauravaram, P., & Kelsey, J. (2007). Cryptanalysis of a class of cryptographic hash functions. Cryptology EPrint Archive.

[36]. Toz, D. (2013). Cryptanalysis of Hash Functions (Cryptanalyse van hashfuncties).

CHAPTER 9

Symmetric Encryption Algorithms

In this chapter, we will discuss two types of symmetric encryption algorithms, Data Encryption Standard (DES) and Advanced Encryption Standard (AES). The goal of this chapter is to provide a starting point for implementing such algorithms from scratch without using third-party libraries, by using the theoretical concepts and translating them into a practical application using Java.

Data Encryption Standard

This section will describe DES as it was published by the National Institute for Standards and Technology (NIST) in publication FIPS PUB 46-3 [5]. DES is a block cipher that is included in the symmetric encryption systems category, as it uses just one secret key for the encryption and for the decryption. At this point, DES should not be used in real applications because it has been cracked, and the literature is rich with different cryptanalysis attacks [4].

DES is an algorithm that has its origin in the LUCIFER algorithm, also known as the Feistel block cipher algorithm. It was developed by Horst Feistel at IBM, a brilliant researcher. Its security was based on the secret key represented in blocks of 128 bits and another block also represented on 128 bits. Another important aspect that is related to its design is the fact that the algorithm uses 16 rounds within the Feistel structure. Each round has its own key.

S. L. Nita and M. I. Mihailescu, *Cryptography and Cryptanalysis in Java*, https://doi.org/10.1007/978-1-4842-8105-5_9

In 1976, it was recognized and adopted as a Federal Encryption Standard. In 1995, an advanced version of the DES algorithm, known as Triple DES (for short, 3DES or TDES), was proposed. Starting with 2002, AES (which will be discussed later) was released, and it was used to replace the DES encryption algorithm as a recognized and adopted standard. Its official name is the Triple Data Encryption Algorithm (for short, TDEA or 3DEA).

TDEA represents another symmetric-key block algorithm that is based on the DES algorithm. For each data block, TDEA will apply DES three times. The size of the block is set to 64 bits, and the sizes of the keys are 56 bits, 112 bits, and 168 bits. Each key is represented as 1, 2, and 3, as shown in Figure 9-1. The rounds are equivalent to the ones from DES, with 48 rounds in total, which means 16 rounds for each key ($16 \times 3 = 48$ rounds).

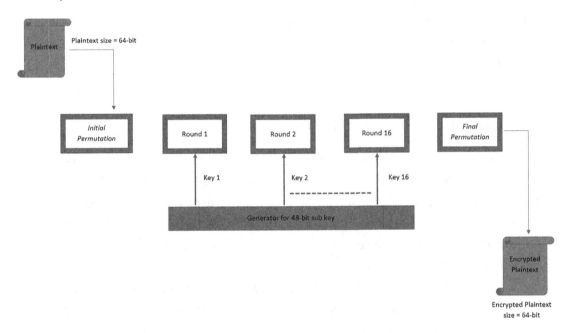

Figure 9-1. *Encryption process for 64-bit plaintext*

According to Figure 9-1, DES will encrypt the data using the first key (Key 1), and the decryption will be done using the second key (Key 2). Further, the encryption is performed again by using the third key (Key 3). There is another version of the algorithm in which only the first and third keys are used, and the keys are the same. Some applications still use this algorithm, but it is treated as a legacy algorithm.

Starting with Figure 9-2, you can see how the encryption and decryption processes are performed within TDEA.

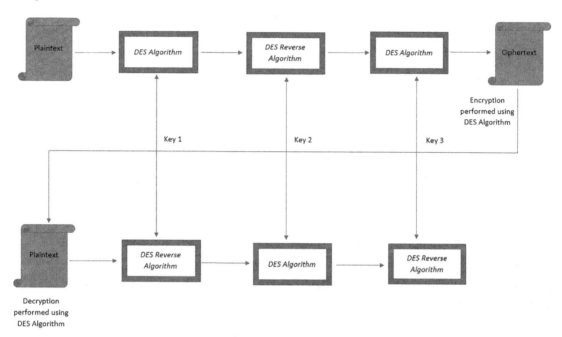

Figure 9-2. *Encryption and decryption process using TDEA*

At this point, we will not focus on the mathematical background, and we will proceed further with the implementation process. We strongly recommend following a theoretical course in cryptography and information security to go deep and understand what is happening during the implementation process. To achieve this goal, you'll find good references at the end of the chapter that will provide the best knowledge for a cryptography mathematical apparatus.

The Generation of Keys

The DES algorithm will compute 16 rounds of the encryption process for each of the rounds a genuine and unique key is generated. To continue with the next steps, it is important to understand how the bits representation of the plaintext are labeled, from 1 to 64. The most significant bit is represented by 1, and the least significant bit is represented by 64.

Let's take a look at the process of generating the keys. For this process, you need the PC-1 and PC-2 permutations; see Table 9-1.

Table 9-1. *Permutation PC-1 [5]*

57	49	41	33	25	17	9
1	58	50	42	34	26	18
10	2	59	51	43	35	27
19	11	3	60	52	44	36
63	55	47	39	31	23	15
7	62	54	46	38	30	22
14	6	61	53	45	37	29
.21	13	5	28	20	12	4

To transform the permutation PC-1 or PC-2 from Table 9-1 or Table 9-2 in a Java expression, it is sufficient to declare a variable, for example, `int pc_1[56]` and `int pc2[48]`, respectively, and then initialize them properly as an array with a size set to 56 and 48, respectively.

Table 9-2. *Permutation PC-2 [5]*

14	17	11	24	1	5
3	28	15	6	21	10
23	19	12	4	26	8
16	7	27	20	13	2
41	52	31	37	47	55
30	40	51	45	33	48
44	49	39	56	34	53
46	42	50	36	29	32

Encryption and Decryption Process

To perform the encryption process, the following steps should be followed to preserve the logic during the implementation process:

- Input

  - Data ← plaintext

  - Size ←64 – bit size

- The plaintext is converted to an initial permutation (IP) function.

- The IP will divide the plaintext into two equal-sized permuted blocks. In some documentation, these permuted blocks are called the Left Plain Text (LPT) and the Right Plain Text (RPT).

- Using the two blocks generated, LPT and RPT, the 16 rounds related to the encryption process are performed. The steps behind the 16 rounds are as follows:

  - Step 1: Key transformation

  - Step 2: Expansion permutation

  - Step 3: S-box permutation

  - Step 4: P-box permutation

  - Step 5: XOR

  - Step 6: Swap

- Once the encryption process is done, the two blocks (LPT and RPT) are rejoined.

- In the end, we get the encrypted version of the plaintext that will have a size of 64 bits.

The decryption process is based on using the same algorithm but in reverse order.

Another important aspect that is taken into consideration when implementing the DES algorithm is its modes of operation.

Operation Modes for DES

DES contains five modes of operation from which a developer can choose during the implementation process.

- *Cipher Block Chaining (CBC)*: In this mode, each block is strictly connected with the previous node. It is using an initialization vector (IV).

- *Electronic Codebook (ECB)*: For each block with a 64-bit size, the encryption and decryption are applied independently.

- *Cipher Feedback (CFB)*: Once the ciphertext is obtained, it will be considered as the input for the algorithm. A pseudorandom output will be obtained from this model. The resulted output is computed using an XOR operation of the plaintext. Based on the XOR operation, the encryption is computed, and its result is passed further to the next operation.

- *Counter (CTR)*: For each of the blocks from the plaintext, the XOR operation is performed with the help of an encrypted counter. The next step is quite tricky as it needs to be incremented sequentially for each of the blocks.

- *Output Feedback (OFB)*: There are some similarities with the CFB mode. The single difference is represented by the input of the encryption algorithm, which is the output that precedes DES.

At this moment, we can say that we have covered all the elements and main concepts to perform the implementation process.

In Listing 9-1 we will provide a basic implementation for the DES algorithm in which we will show all the steps.

Listing 9-1. DES Algorithm (Encryption and Decryption)

```
1    import java.io.*;
2    import java.security.spec.*;
3    import javax.crypto.*;
4    import javax.crypto.spec.*;
5
6
7    public class DES
8    {
9         //instance for encryption
10        private static Cipher encOp;
11
12        //instance for decryption
13        private static Cipher decOp;
14
15        //path for the file that will be encrypted
```

```
16          private static final String textFileToEncrypt = "D:/
            apressFile.txt";
17
18          //path for the encrypted file
19          private static final String encFile = "D:/apress_enc.txt";
20
21          //path for decryptied file
22          private static final String decFile = "D:/apress_dec.txt";
23
24          //vector for initialization
25          private static final byte[] iv = { 25, 38, 15, 43, 59, 92, 66, 73 };
26
27          public static void main(String[] args)
28          {
29              try
30              {
31                  //setting up the key
32                  SecretKey secret_key = KeyGenerator.
                    getInstance("DES").generateKey();
33                  AlgorithmParameterSpec parameters_specs = new
                    IvParameterSpec(iv);
34
35
36                  //specify the encryption mode
37                  encOp = Cipher.getInstance("DES/CBC/PKCS5Padding");
38                  encOp.init(Cipher.ENCRYPT_MODE, secret_key,
                    parameters_specs);
39
40                  //specify the decryption mode
41                  decOp = Cipher.getInstance("DES/CBC/PKCS5Padding");
42                  decOp.init(Cipher.DECRYPT_MODE, secret_key,
                    parameters_specs);
43
44                  //encrypt the file
```

```
45              File forEncryption = new File(textFileToEncrypt);
46              if(!forEncryption.exists())
47                  throw new FileNotFoundException("The file does
                    not exist. Please create it and write the text
                    you wish to encrypt.");
48              ComputeEncOperation(new FileInputStream(textFileTo
                Encrypt), new FileOutputStream(encFile));
49
50              //decrypt the file
51              ComputeDecOperation(new FileInputStream(encFile),
                new FileOutputStream(decFile));
52
53              //show a message for letting the user know the
                situation
54              System.out.println("The files for encryption and
                decryption results have created successfully.");
55          }
56
57          //catch any exception encountered
58          catch (Exception e)
59          {
60              //prints the message (if any) related to exceptions
61              e.printStackTrace();
62          }
63      }
64
65      //write bytes content to the files
66      private static void writeToFileTheBytes(InputStream input,
        OutputStream output) throws IOException
67      {
68          byte[] writeBuffer = new byte[512];
69          int readBytes = 0;
70          while ((readBytes = input.read(writeBuffer)) >= 0)
```

```
71                    {
72                            output.write(writeBuffer, 0, readBytes);
73                    }
74                    //closing the output stream
75                    output.close();
76                    //closing the input stream
77                    input.close();
78            }
79
80            //encryption operation
81            private static void ComputeEncOperation(InputStream inFile,
               OutputStream ouFile) throws IOException
82            {
83                    ouFile = new CipherOutputStream(ouFile, encOp);
84
85                    //write the bytes obtained from the encryption to the file
86                    writeToFileTheBytes(inFile, ouFile);
87            }
88
89            //decryption operation
90            private static void ComputeDecOperation(InputStream inFile,
               OutputStream ouFile) throws IOException
91            {
92                    inFile = new CipherInputStream(inFile, decOp);
93
94                    //write the bytes obtained from the decryption to the file
95                    writeToFileTheBytes(inFile, ouFile);
96            }
97    }
```

Before running the implementation from Listing 9-1, we will need to have a file created at a specified location, such as the one specified in line 16 (see Figure 9-3). If the file does not exist, an exception is thrown with an error message (line 55). The other two files are created automatically.

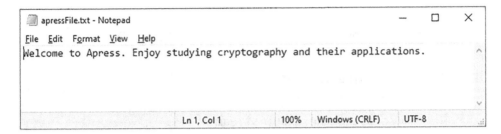

Figure 9-3. *The file and its content that will be encrypted*

Based on the file specified in line 16, when the encryption (see line 100) and decryption operations (see line 110) are performed, two files will be generated: apress_ enc.txt for encryption output (see line 20 and Figure 9-4) and apress_dec.txt for decryption output (see line 23 and Figure 9-5).

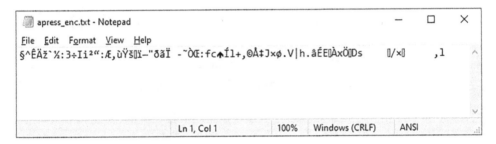

Figure 9-4. *File apress_enc.txt*

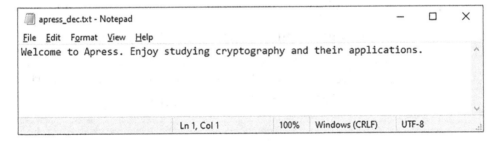

Figure 9-5. *File apress_dec.txt*

Advanced Encryption Standard

The Advanced Encryption Standard represents a symmetric block cipher that provides encryption and decryption operations, published by NIST under the publication FIPS 197 [6]. This section will follow the technical description of AES as it is in [6]. The size of the key accepted is 128, 192, or 256 bits. The same key used for encryption has to be used during the decryption process as well.

An important aspect of AES during the encryption process is that the data is converted into a block, and then the encryption is made using padding operations with blocks of length 128, 192, or 256 bits.

Several advantages can be summarized as follows:

- The encrypted message/file can be decrypted without the key that has been used in the encryption stage. A valid secret key has to be used.

- AES is used in different areas of applications, such as data storage, financial transactions, wireless device communication, and much more.

- It is a secure algorithm, but some cryptanalysis attacks have been launched with success [3].

Also, there are some disadvantages, such as the following:

- The mechanism (mathematical apparatus) behind AES is quite simple. For further reading, I strongly recommend pursuing a cryptography course such as [1, 2].

- AES can represent a serious challenge during the implementation of the software.

Encryption and Decryption Process

During the encryption process, the plaintext will be encrypted using a secret key that is known only by the sender and the receiver of the message.

Within Java, the encryption and decryption are represented by the Java Cryptographic Extension (JCE) platform. JCE offers different packages that can be used by developers during their implementation processes, such as `java.security` and `javax.crypto` (see lines 4–8 from Listing 9-2).

The same process that is used for encryption is used as well for decryption. With the help of the init() method from the Cipher class, the algorithm will be initialized with the public key based on the chosen transformation.

Operation Modes for AES

There are five modes of operation from which a developer can choose during the implementation process.

- *Cipher Block Chaining (CBC)*: In CBC mode the initialization vector is used for getting a better process of the encryption process. To obtain the encryption output, the CBC is using an XOR operation between the initialization vector and plaintext.

- *Electronic Codebook (ECB)*: This is one of the simplest modes of all modes available. The plaintext is divided into blocks, with each block having 128 bits. The same key and algorithm chosen in the beginning are used to encrypt the blocks.

- *Cipher Feedback (CFB)*: The first operation performed by CFB is to perform the encryption of the initialization vector. Once we have the encryption of the initialization vector, an XOR operation with the plaintext is performed, and the encrypted message is obtained. Moving forward, the encrypted message (or block) is encrypted with the next plaintext message (or block).

- *Counter (CTR)*: For each of the blocks from the plaintext, the XOR operation is performed with the help of an encrypted counter. The next step is quite tricky as it needs to be incremented sequentially for each of the blocks.

- *Output Feedback (OFB)*: There are some similarities with the CFB mode. The single difference is represented by the input of the encryption algorithm, which is the output that precedes DES.

- *Galois and Counter Mode (GCM)*: This mode represents an extended version for CTR mode. It has been proposed and introduced by NIST. In GCM mode we have the ciphertext and an authentication label once the encryption process is done.

Based on the elements that we have mentioned, we can say that we have all the elements and main concepts set to perform with the implementation process.

In Listing 9-2 we will provide a basic implementation for the AES algorithm in which we will emphasize all the steps . Figure 9-6 shows the result.

Listing 9-2. Implementation of AES Algorithm

```
1    import java.util.*;
2    import java.security.spec.*;
3    import javax.crypto.*;
4    import javax.crypto.spec.*;
5    import java.nio.charset.*;
6
7
8    public class AES
9    {
10         private static byte[] initialization_vector = {54, 34, 7, 3, 23,
           78, 31, 68, 32, 40, 96, 43, 23, 54, 23, 76};
11         private static String aes_secretKey = "cryptoApress";
12         private static String aes_salt = "apress";
13
14          public static String Encrypt(String plain_message)
15          {
16              try
17              {
18                      IvParameterSpec initializationVectorSpecs = new
                        IvParameterSpec(initialization_vector);
19
20                      //the container for the secret key
21                      SecretKeyFactory secretKeyContainer = SecretKey
                        Factory.getInstance("PBKDF2WithHmacSHA256");
22
23                      //specification parameters (secret key, salt value,
                        iterations, key length)
24                      KeySpec specificationsParameters = new PBEKeySpec
                        (aes_secretKey.toCharArray(),aes_salt.getBytes(),
                        65536, 256);
```

```
25
26                    //generate the secret key based on the parameters
                      set above
27                    SecretKey temporarySecretKey = secretKeyContainer.
                      generateSecret(specificationsParameters);
28
29                    //align the secret key with the AES algorithm
30                    SecretKeySpec crypto_key = new SecretKeySpec
                      (temporarySecretKey.getEncoded(), "AES");
31
32                    //set the algorithm (e.g., AES) and its mode
                      together with its padding type
33                    Cipher aesAlgorithm = Cipher.getInstance
                      ("AES/CBC/PKCS5Padding");
34                    aesAlgorithm.init(Cipher.ENCRYPT_MODE, crypto_key,
                      initializationVectorSpecs);
35
36                    //get the encrypted version
37                    return Base64.getEncoder().encodeToString
                      (aesAlgorithm.doFinal(plain _message.
                      getBytes(StandardCharsets.UTF_8)));
38                }
39            catch (Exception e)
40            {
41                    System.out.println("During the encryption process,
                      the following error(s) occured: " + e.toString());
42            }
43            return null;
44      }
45
46      public static String Decrypt(String encrypted_message)
47          {
48          try
49              {
```

```
50          IvParameterSpec initializationVectorSpecs = new
            IvParameterSpec(initialization_vector);
51
52          //the container for the secret key
53          SecretKeyFactory secretKeyContainer =
            SecretKeyFactory.getInstance("PBKDF2WithHmacSHA256");
54
55          //specification parameters (secret key, salt value,
            iterations, key length)
56          KeySpec specificationsParameters = new PBEKeySpec
            (aes_secretKey.toCharArray(),aes_salt.getBytes(),
            65536, 256);
57
58          //generate the secret key based on the parameters
            set above
59          SecretKey temporarySecretKey = secretKeyContainer.
            generateSecret(specificationsParameters);
60
61          //align the secret key with the AES algorithm
62          SecretKeySpec secretKey = new SecretKeySpec
            (temporarySecretKey.getEncoded(), "AES");
63
64          //set the algorithm (e.g., AES) and its mode together
            with its padding type
65          Cipher cipher = Cipher.getInstance("AES/CBC/
            PKCS5PADDING");
66          cipher.init(Cipher.DECRYPT_MODE, secretKey,
            initializationVectorSpecs);
67
68          //get the decrypted value
69          return new String(cipher.doFinal(Base64.getDecoder().
            decode(encrypted_message)));
70      }
71  catch (Exception e)
```

```
72              {
73                      System.out.println("Error occured during decryption: "
                        + e.toString());
74              }
75              return null;
76          }
77
78
79      public static void main(String[] args)
80      {
81          //set the message that we want to encrypt
82         String originalval = "Welcome to Apress. Enjoy learning
           cryptography";
83
84         //perform the encryption
85         String encryptedval = Encrypt(originalval);
86
87         //perform the decryption
88         String decryptedval = Decrypt(encryptedval);
89
90         //show some messages
91         System.out.println("Plaintext used for encryption and
           decryption -> " + originalval);
92         System.out.println("The encryption is -> " + encryptedval);
93         System.out.println("The decryption is -> " + decryptedval);
94      }
95  }
```

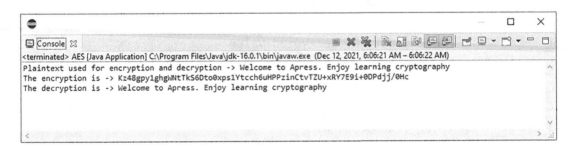

Figure 9-6. *The output of encryption/decryption with AES*

Before running the implementation from Listing 9-2, we need to set the initialization vector, which can be set manually (see line 10); the secret key (see line 13); and the salt value (see line 15). As a best practice, we strongly recommend that the initialization vector be generated randomly without hard-coding it. This method is used only for the example and educational purposes.

Conclusion

In this chapter, we discussed symmetric encryption algorithms such as Data Encryption Standard and Advanced Encryption Standard. You now should understand the main differences, advantages, and disadvantages of DES and AES, as well as a short summary of the modes used for each of the algorithms.

The examples provided were straightforward; we used a common and basic implementation and explained the mode in which they should be implemented. The implementations that we have provided are offered for educational purposes, and there is no guarantee that in practice they will offer a maximum benefit, especially in the case of the AES algorithm, which is quite sensitive about the process of implementation.

Lastly, Figure 9-7 summarizes the encryption process between DES and AES, as shown in Figure 9-7.

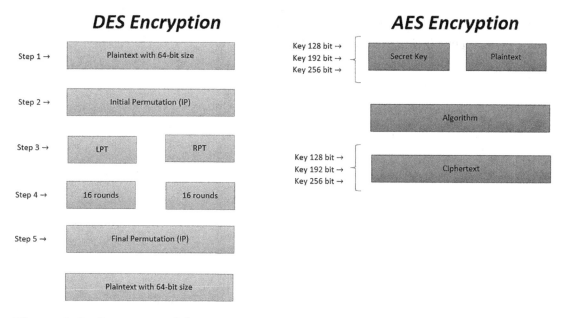

Figure 9-7. *Summary of the encryption process between DES and AES*

References

[1].   Stinson, Douglas R., and Maura B. Paterson. Cryptography: Theory and Practice. Fourth edition, CRC Press, Taylor & Francis Group, 2019.

[2].   Menezes, A. J., et al. Handbook of Applied Cryptography. CRC Press, 1997.

[3].   Dobbertin H., Knudsen L., Robshaw M. (2005) The Cryptanalysis of the AES – A Brief Survey. In: Dobbertin H., Rijmen V., Sowa A. (eds) Advanced Encryption Standard – AES. AES 2004. Lecture Notes in Computer Science, vol 3373. Springer, Berlin, Heidelberg. https://doi.org/10.1007/11506447_1.

[4].   Biham, E., Shamir, A. Differential cryptanalysis of DES-like cryptosystems. *J. Cryptology* **4**, 3–72 (1991). https://doi.org/10.1007/BF00630563

[5].   FIPS PUB 46-3. Available online: https://csrc.nist.gov/csrc/media/publications/fips/46/3/archive/1999-10-25/documents/fips46-3.pdf

[6].   FIPS PUB 197. Available online: https://nvlpubs.nist.gov/nistpubs/FIPS/NIST.FIPS.197.pdf

CHAPTER 10

Asymmetric Encryption Schemes

Another name for asymmetric encryption is *public-key cryptography* (PKC). This name comes from the fact that the cryptosystems in this category are using a pair of keys in their algorithms. Specifically, a *public key* is used by the encryption algorithm, and a *private (secret) key* is used by the decryption algorithm. In classical cryptography, the purpose of each type of key is well established, but in modern cryptography, these types of keys can be used for additional purposes. For example, another type of secret key can be a master key, used to encrypt and decrypt the keys of a specific cryptosystem.

While in symmetric encryption schemes the cryptosystems use mathematical concepts such as substitutions and permutations (learn about the AES and DES algorithms in Chapter 9), asymmetric encryption has asymmetric functions as its foundations. The authors who introduced the concept of asymmetric encryption are Diffie and Hellman in their paper "New directions in cryptography" [1]. In this paper, the authors explained what asymmetric encryption is and what the characteristics of such a cryptosystem are. In addition, they showed how the secret keys can be securely distributed over an insecure channel between the sender and the receiver. This new concept was welcomed by the scientific community, and many encryption systems that follow the rules from [1] were proposed in the meantime. The following are the requirements presented in [1] that need to be accomplished by the cryptosystems:

- The receiver (which is the owner of the secret key) should be able to use a computationally efficient algorithm to generate the pair of keys.

- A pair of keys is used. The public key is used to encrypt the plain messages, and the secret key is used to decrypt the encrypted messages.

© Stefania Loredana Nita and Marius Iulian Mihailescu 2022
S. L. Nita and M. I. Mihailescu, *Cryptography and Cryptanalysis in Java*,
https://doi.org/10.1007/978-1-4842-8105-5_10

- The sender should be able to use a computationally efficient algorithm to encrypt the plain text (message) into the encrypted text.

- The receiver should be able to use a computationally efficient algorithm to recover the plain text (message) from the encrypted text.

- The secret key should be independent of the public key; therefore, the public key cannot be used as input for the secret key generation.

- The plain text (message) cannot be computed if one knows the public key and the encoded message.

Generally, an encryption system that uses a pair of keys as described works as follows between the sender (S) and the receiver (R):

1. S and R generate a pair of public and private keys, making the public key available to everyone, while the private key is kept secret by each party.

2. To send a message, S uses the public key of R and encrypts the message using the encryption algorithm.

3. To recover the message, R uses its private key and the decryption algorithm to recover the message.

Some examples of well-known public-key encryption systems are RSA [2], ElGamal [3], and Merkle-Hellman [4], which will be explained individually in the remainder of this chapter. Asymmetric encryption is important because it is used in different applications in cryptography, such as digital certificates and signatures, protocols (encryption, multiparty computation, zero-knowledge), etc. Types of encryption schemes include asymmetric encryption, homomorphic encryption, searchable encryption, predicate encryption, functional encryption, etc., while branches of public-key cryptography are integer factorization cryptography, elliptic-curve cryptography, lattice-based cryptography, learning and ring-learning with errors, multivariate cryptography, code-based cryptography, etc.

Sometimes, symmetric cryptosystems are faster than asymmetric ones, although they may have the same security level. Usually, in these cases, the plain messages are encrypted using the symmetric encryption scheme, and the secret key is encrypted using the asymmetric encryption scheme and is sent between the sender and receiver over an insecure channel.

RSA

A well-known and commonly used public-key cryptosystem is RSA [2], which was introduced in 1978 and named after its authors: Ron Rivest, Adi Shamir, and Len Adleman. The hardness assumption for RSA is integer factorization, which means that for two prime numbers that are large enough, their product is *hard* to be factorized by a computing machine. What specifically does hard mean? It means that the problem to be solved requires more than polynomial time.

In the remainder of this section, we will present the algorithms of the RSA cryptosystem [2] and then their implementation in Java.

The first algorithm is the *key generation* algorithm, and it works as follows:

1. Generate two distinct large prime numbers p, q.

2. Determine $n = pq$.

3. Determine $\phi(n) = (p - 1)(q - 1)$. The ϕ function is called Euler's totient function.

4. Generate the integer value e, with $1 < e < \phi(n)$ and $\gcd(e, \phi(n)) = 1$.

5. Determine $d \equiv e^{-1}(mod\ \phi(n))$.

The public key of the cryptosystem is $k_p = (n, e)$, and the private (secret) key is $k_s = (p, q, \phi(n), d)$; thus, in the encryption algorithm, the values n and e will be used, and in the decryption algorithm the values p, q, $\phi(n)$, and d will be used.

The next algorithm is the *encryption algorithm*, which encrypts the plain message using the following formula. The plain message is taken as an integer value m, $0 \leq m < n$, and the encrypted value obtained is c:

$$m^e \equiv c\ (mod\ n)$$

The last algorithm is the *decryption algorithm*, which recovers the plain message m from the encrypted value c, using the following formula:

$$c^d \equiv \left(m^e\right)^d \equiv m\ (mod\ n)$$

Listing 10-1 presents the RSA algorithm in Java.

Listing 10-1. Implementation of the RSA Cryptosystem

```
1    import java.io.BufferedReader;
2    import java.io.IOException;
3    import java.io.InputStreamReader;
4    import java.math.BigInteger;
5    import java.security.SecureRandom;
6
7
8    public class RSA {
9
10       private BigInteger prime_p, prime_q, val_n, phi_n, val_e, val_d;
11       private BigInteger one = BigInteger.ONE;
12       private int maximumLength = 1024;
13       private SecureRandom random;
14
15       public void KeyGeneration() {
16            random = new SecureRandom();
17            prime_p = BigInteger.probablePrime(maximumLength, random);
18            prime_q = BigInteger.probablePrime(maximumLength, random);
19            val_n = prime_p.multiply(prime_q);
20            phi_n = prime_p.subtract(one).multiply(prime_q.
                 subtract(one));
21            val_e = BigInteger.probablePrime(maximumLength, random);
22            do {
23                 val_e = BigInteger.probablePrime(maximumLength, random);
24            }
25            while (phi_n.gcd(val_e).compareTo(one) < 0 && val_e.
                 compareTo(phi_n) > 0);
26            val_d = val_e.modInverse(phi_n);
27       }
28
29
```

```
30      public byte[] Encryption(byte[] plainMessage, BigInteger e,
        BigInteger n)
31      {
32          BigInteger encryptedMessage = (new BigInteger(plainMessage)).
            modPow(e, n);
33          return encryptedMessage.toByteArray();
34      }
35
36      public byte[] Decryption(byte[] encryptedMessage, BigInteger d,
        BigInteger n)
37      {
38          BigInteger decryptedMessage = (new BigInteger
            (encryptedMessage)).modPow(d, n);
39          return decryptedMessage.toByteArray();
40      }
41
42       public static void main (String [] arguments) throws IOException
43      {
44          RSA rsa = new RSA();
45          rsa.KeyGeneration();
46
47          BufferedReader d = new BufferedReader(new
            InputStreamReader(System.in));
48          String plainMessage;
49          System.out.print("Type the plain message: ");
50          plainMessage = d.readLine();
51
52          System.out.println("\nEncrypting the message... ");
53          byte[] encryptedMessage = rsa.Encryption(plainMessage.
            getBytes(),
54          rsa.val_e, rsa.val_n);
55          System.out.println("Encrypted message [bytes]: " + byteToStrin
            g(encryptedMessage));
56          System.out.println("Encrypted message [text]: " + new
            String(encryptedMessage));
```

```
57
58            System.out.println("\nDecrypting the message... ");
59            byte[] decryptedMessage = rsa.Decryption(encryptedMessage,
              rsa.val_d, rsa.val_n);
60            System.out.println("Decrypted message [bytes]: " + byteTo
              String(decryptedMessage));
61            System.out.println("Decrypted message [text]: " + new
              String(decryptedMessage));
62        }
63
64
65        private static String byteToString(byte[] byteArray)
66        {
67            String recoveredStrig = "";
68            for (byte byteVal : byteArray)
69            {
70                    recoveredStrig += Byte.toString(byteVal);
71            }
72            return recoveredStrig;
73        }
74    }
```

The code from Listing 10-1 follows the steps of the RSA algorithms presented in this section. In lines 1–5, the necessary packages were imported, as follows: BufferedReader and InputStreamReader are used for plain messages on line 50. The IOException is necessary for cases where the input introduced by the user is not correct and makes that main function to throw an IOException in this case. The BigInteger package is used to work with big integers in Java. As shown, BigInteger is a Java class (and not a third-party one). SecureRandom is used to generate pseudorandom numbers securely (see Chapter 7 for details). In lines 10–13, the variables used in the RSA cryptosystem are declared (see the descriptions of the algorithms from the previous listing), with the prime_ prefix showing that the parameter should be a prime number and val_ showing that the parameter is an integer value. The KeyGeneration, Encryption, and Decryption functions just follow the mathematical operations of the RSA cryptosystem for each algorithm. Note that BigInteger already implements the mathematical operations (including prime number generation), which makes it easy to use. The encryption function has as input the plain message and the public key, while the decryption function has as input the encrypted message and the private key. In lines 44–60 the plain message is read from the keyboard; then it is encrypted, and the result is decrypted. All this information is displayed in the console. The function `byteToString` is used just to print the byte array in a neat form, converting it to `String`. The size for the keys is 1024 (given by the value of `maximumLength`); in practice, the key size of RSA is 512, 1024, or even 2048. Figure 10-1 shows the output of Listing 10-1. Note that the values obtained in Figure 10-1 can be different every time because pseudorandom numbers are used. However, the plain message should be the same as the decrypted message every time.

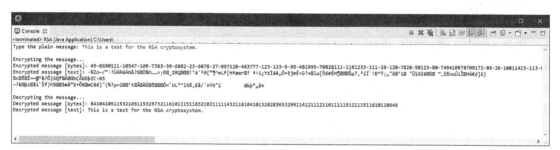

Figure 10-1. *The output for the RSA cryptosystem*

As the implementation from Listing 10-1 uses `SecureRandom` and `BigIntegers`, it is not a naïve implementation, as it would be if simple data types (such as `int`) were used. However, the implementation of RSA can be even more advanced. Such an example can be found in [9], where the author uses `javax.crypto` and `X509EncodedKeySpec` to encode the public key. Regarding cryptanalysis, for example, a Pollard $p-1$ attack [11] or continued fraction attack [12] can be launched against the RSA encryption system. Because the public key is known by anyone, both attacks can find the private key if the values for d are small. A good implementation of these attacks is provided in [10].

Because of its construction, RSA is one of the most important public-key encryption systems. Based on it, a new hardness assumption was derived from the integer factorization, called the *RSA problem*, which is the foundation for subsequent cryptosystems. The RSA hardness assumption lies in the difficulty of computing the secret key by knowing the public key. Until this moment, there is no known efficient algorithm that solves this problem for a key size larger than 1,024 bits.

Another interesting feature of RSA is the homomorphic property of the multiplication operation. In the last decade or so, homomorphic encryption is a key branch in cryptography, because it allows (partially or totally, but limited) to make computations directly with the encrypted text, and the decryption of the result is the same as the result of the equivalent operations (via a homomorphism) applied on the plain text. Chapter 14 will discuss homomorphic encryption in more detail. For the RSA cryptosystem, the homomorphic operation is the following multiplication:

$$Enc(m_1) \bullet Enc(m_2) = m_1^e \bullet m_2^e \ (mod \ n) = (m_1 \bullet m_2)^e \ (mod \ n) = Enc(m_1 \bullet m_2)$$

ElGamal

Another notable example of public-key cryptosystems is ElGamal [3], whose theoretical foundations lie in the cyclic groups and the hardness assumption is the discrete logarithm problem (DLP). The specific DLPs for the ElGamal encryption system are the Diffie-Hellman assumption and the decisional Diffie-Hellman assumption (more technical details about DLP and Diffie-Hellman assumptions can be found in [5] and [6]). In this section, we present the ElGamal algorithms [3] and then show the implementation in Java.

The first algorithm is the *key generation algorithm*, and it works as follows:

1. Pick a prime number q and pick a cyclic group G (i.e., an element $g \in G$ should generate the entire group G) with the identity element $e \in G$.

2. Take an integer value from the interval $x \in \{1, ..., q - 1\}$.

3. Compute the value $h = g^x$.

After running the key generation algorithm, the public key is $k_p = (G, q, g, h)$, and the private/secret key is $k_s = x$.

The second algorithm is the *encryption algorithm*, which is used to encrypt the plain message $m \in G$, and it is proceeds as follows:

1. Pick randomly a value $y \in \{1, ...q - 1\}$.

2. Compute $s = h^y$.

3. Compute $c_1 = g^y$.

4. Compute $c_2 = m \bullet s$.

The encryption of the plain message m is the pair $c = (c_1, c_2)$.

The third algorithm is the decryption algorithm used to decrypt the encrypted message $c = (c_1, c_2)$ and to recover the plain message. It has the following steps:

1. Compute $s = c_1^x$.

2. Compute s^{-1}, the inverse of s.

3. Compute $m = c_2 \bullet s^{-1}$.

The cryptosystem is correct because the plain message is always recovered: $c_2 \bullet s^{-1} = (m \bullet s) \bullet s^{-1} = m$. An important technical aspect is that all operations are made within the group G; thus, all obtained results are elements of G.

The value s from the third step of the encryption algorithm is called the *shared secret*. This value is shared because, with some computations, it can be easily verified that it is the same as the value from the first step of the decryption $c_1^x = \left(g^y\right)^x = g^{xy} = h^y$. So, the sender and the receiver share the same value h^y.

The parameters of the system should be wisely chosen because, by taking a look at the encryption algorithm, we can see that an attacker knows $c = (c_1, c_2)$ and succeeds to recover m; then the shared secret can be computed as $c_2 \bullet m^{-1} = s$. In practice, to avoid this vulnerability, when a new message should be encrypted, a new value for y should be chosen. Listing 10-2 presents the Java implementation of the ElGamal cryptosystem.

Listing 10-2. ElGamal Implementation

```
1    import java.io.BufferedReader;
2    import java.io.IOException;
3    import java.io.InputStreamReader;
4    import java.math.BigInteger;
5    import java.security.SecureRandom;
6
7    public class ElGamal {
8
9            private BigInteger prime_q, val_x, val_g, val_h;
10           private byte[] c1, c2;
11           private BigInteger one = BigInteger.ONE;
12           private BigInteger two = new BigInteger("2");
13           private int maximumLength = 1024;
```

```
14          private SecureRandom random;
15
16          public void KeyGeneration() {
17                  random = new SecureRandom();
18
19                  prime_q = BigInteger.probablePrime(maximumLength, random);
20
21                  do {
22                          val_x = BigInteger.probablePrime(maximumLength, random);
23                  }
24                  while (val_x.compareTo(prime_q.subtract(one)) >= 0);
25
26                  BigInteger p2 = prime_q.subtract(one);
27              p2 = p2.divide(two);
28
29              // take a generator g of the group
30              val_g = new BigInteger(maximumLength, random);
31              val_g = val_g.mod(prime_q);
32              while(val_g.modPow(p2,prime_q).compareTo(prime_q.
                subtract(one)) != 0)
33              {
34                      val_g = new BigInteger(maximumLength, random);
35                      val_g = val_g.mod(prime_q);
36              }
37
38              val_h = val_g.modPow(val_x, prime_q);
39          }
40
41
42      public void Encryption(byte[] plainMessage)
43      {
44              BigInteger y, s;
45              do {
46                      y = new BigInteger(maximumLength, random);
```

```
47              }
48              while (y.compareTo(val_h) >= 0);
49
50          s = val_h.modPow(y, prime_q);
51          c1 = (val_g.modPow(y, prime_q)).toByteArray();
52          c2 = (new BigInteger(plainMessage).multiply(s)).
            toByteArray();
53
54          System.out.println("Encrypted message [bytes]: " +
            byteToString(c1) + byteToString(c2));
55      System.out.println("Encrypted message [text]: " + new
        String(c1) + new String(c2));
56
57      }
58
59      public void Decryption()
60      {
61          BigInteger s = new BigInteger(c1).modPow(val_x, prime_q);
62          BigInteger invS = s.modInverse(prime_q);
63          BigInteger m = invS.multiply(new BigInteger(c2)).
            mod(prime_q);
64
65          System.out.println("\nDecrypted message [bytes]: " + byteTo
            String(m.toByteArray()) + byteToString(m.toByteArray()));
66      System.out.println("Decrypted message [text]: " + new
        String(m.toByteArray()));
67      }
68
69
70      public static void main(String[] args) throws IOException {
71          ElGamal elGamal = new ElGamal();
72          elGamal.KeyGeneration();
73
74      BufferedReader d = new BufferedReader(new InputStreamReader
        (System.in));
```

```
75          String plainMessage;
76          System.out.print("Type the plain message: ");
77          plainMessage = d.readLine();
78
79          System.out.println("\nEncrypting the message... ");
80          elGamal.Encryption(plainMessage.getBytes());
81          elGamal.Decryption();
82        }
83
84      private static String byteToString(byte[] byteArray)
85        {
86          String recoveredStrig = "";
87          for (byte byteVal : byteArray)
88          {
89              recoveredStrig += Byte.toString(byteVal);
90          }
91          return recoveredStrig;
92        }
93    }
```

Similarly, with RSA, the ElGamal implementation follows the algorithms described earlier. Figure 10-2 shows the result. Note that the values obtained in Figure 10-2 can be different every time because pseudorandom numbers are used. However, the plain message should be the same as the decrypted message every time. Examples of attacks that can be launched against the ElGamal cryptosystem are brute-force attacks, meet-in-the-middle attacks, and two-table attacks. A good reference for these attacks on ElGamal is [13].

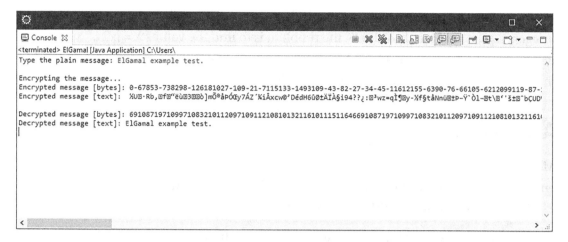

Figure 10-2. *The result of the ElGamal implementation*

ElGamal is one of the best examples of an asymmetric encryption scheme. This cryptosystem also has homomorphic properties as follows:

$$Enc(m_1) \bullet Enc(m_2) = \left(g^{y_1}, m_1 \bullet h^{y_1}\right) \bullet \left(g^{y_2}, m_2 \bullet h^{y_2}\right) = \left(g^{y_1} \bullet g^{y_2}, m_1 \bullet h^{y_1} \bullet m_2 \bullet h^{y_2}\right)$$
$$= \left(g^{y_1 y_2}, (m_1 m_2) \bullet h^{y_1 y_2}\right) = Enc(m_1 \bullet m_2)$$

Merkle-Hellman

Another example of asymmetric encryption is the Merkle-Hellman cryptosystem [4]. Although it has been proven that it is insecure [8], it is interesting from the technical approach used, specifically the knapsack approach, presented in the next section.

The Knapsack Approach

In combinatorial optimization, the knapsack problem is a common problem. The input for the problem consists of a knapsack characterized by the maximum weight that it supports and a collection of objects characterized by the weight and profit for each object. It is requested to fill the knapsack with objects of the set without exceeding the maximum weight, but by obtaining a maximum profit. However, there are two versions of the problem: the discrete version (works only with whole objects) and the continuous version (may work with fractions of objects).

Technically, the knapsack problem can be formulated as follows: given a set of integers $A = \{a_1, ...a_n\}$ and a value S, the problem is to find the values $X = \{x_1, ..., x_n\}$, $x_i \in [0, 1]$, such that $S = a_1 x_1 + \cdots + a_n x_n$. The work units in cryptography are bits; therefore, the possible values for x_i will become either 0 or 1.

Algorithms

Merkle-Hellman's cryptosystem has the general three algorithms: key generation, encryption, and decryption, which are presented in this section [7].

The *key generation algorithm* has the following steps:

1. Pick an integer value n as the dimension of the block.

2. Pick randomly n positive integer values $W = \{w_1, ..., w_n\}$ that fulfill

    the condition $w_k > \displaystyle\sum_{i=1}^{k-1} w_i, 1 < k \leq n$.

3. Pick randomly an integer q with $q > \displaystyle\sum_{i=1}^{n} w_i$.

4. Pick randomly an integer r with $\gcd(r, q) = 1$.

5. Calculate the elements of the tuple $B = (b_1, ..., b_n)$, where
    $b_i = r w_i \bmod q$.

The public key of the cryptosystem is $k_p = B$, and the secret key is $k_s = (W, q, r)$.

The *encryption algorithm* has one step, and the encrypted value c is obtained using the following formula:

$$c = \sum_{i=1}^{n} m_i b_i$$

where m_1 is the most significant bit and m_i is a bit of the message m with the length n.

The *decryption algorithm* has the following steps for recovering the plain message m from the encrypted value c:

1. Calculate the inverse r' of the value r, using the formula
    $r' = r^{-1} (\bmod q)$.

2. Calculate the value $c' = cr' (\bmod q)$.

3. Pick a subset $X' = \{x_1, ...x_k\}$ for which the elements fulfill the condition $c' = \sum_{i=1}^{k} w_i x_i$.

4. The plain message is recovered as $m = \sum_{i=1}^{k} 2^{n-x_i}$.

The third step of the decryption algorithm represents a problem called the *subset sum problem* for the value c'. This problem can be used by applying a greedy algorithm.

Conclusion

The chapter explained what asymmetric encryption is and whether it is different from symmetric encryption. The three of the most important asymmetric encryption schemes were presented and then implemented: RSA, ElGamal, and Merkle-Hellman.

References

[1]. Diffie, W., & Hellman, M. (1976). New directions in cryptography. IEEE transactions on Information Theory, 22(6), 644-654.

[2]. Rivest, R. L., Shamir, A., & Adleman, L. (1978). A method for obtaining digital signatures and public-key cryptosystems. Communications of the ACM, 21(2), 120-126.

[3]. ElGamal, T. (1985). A public key cryptosystem and a signature scheme based on discrete logarithms. IEEE transactions on information theory, 31(4), 469-472.

[4]. Merkle, R., & Hellman, M. (1978). Hiding information and signatures in trapdoor knapsacks. IEEE transactions on Information Theory, 24(5), 525-530.

[5]. McCurley, K. S. (1990). The discrete logarithm problem. In Proc. of Symp. in Applied Math (Vol. 42, pp. 49-74).

[6]. Bao, F., Deng, R. H., & Zhu, H. (2003, October). Variations of diffie-hellman problem. In International conference on information and communications security (pp. 301-312). Springer, Berlin, Heidelberg.

[7].   Merkle, R., & Hellman, M. (1978). Hiding information and
signatures in trapdoor knapsacks. IEEE transactions on
Information Theory, 24(5), 525–530.

[8].   Shamir, A. (1982, November). A polynomial time algorithm for
breaking the basic Merkle-Hellman cryptosystem. In 23rd Annual
Symposium on Foundations of Computer Science (sfcs 1982)
(pp. 145–152). IEEE.

[9].   RSA in Java, `https://www.baeldung.com/java-rsa`

[10].   RSA-Encryption-Attacks, `https://github.com/toknhawaiian/
RSA-Encryption-Attacks`

[11].   Lydia, M. S., Budiman, M. A., & Rachmawati, D. (2020). On using
Pollard's p-1 Algorithm to Factor RPrime RSA Modulus.

[12].   Dujella, A. (2004). Continued fractions and RSA with small secret
exponent. arXiv preprint cs/0402052.

[13].   Allen, B. (2008). Implementing several attacks on plain ElGamal
encryption. Iowa State University.

CHAPTER 11

Signature Schemes

In the real world, when a message needs to be authenticated by individual, it is written on paper, and then the individual signs the message. Therefore, it is known that this specific message belongs to a specific person. In the digital world, things can happen similarly: a piece of data (message) is created by an individual, and then the message is signed digitally with a specific signature by that individual.

Digital signature schemes are public-key schemes that authenticate a message and check its integrity. The receiver and any third party may check individually the bond between the received message and the sender's digital signature. In the digital signatures schemes, the sender owns a secret key based on which the signature (which is public) is generated. An important aspect is that the sender cannot reject the message once the message is digitally signed by the sender. This property is called *nonrepudiation*, and it is crucial in digital signature schemes. Besides nonrepudiation, other important properties are data integrity and message authentication.

Although they belong to public-key cryptography, the signature schemes have a slightly different form than regular encryption schemes. A digital signature scheme has the following algorithms:

- *Key generation*: This algorithm generates the pair of keys, namely, the secret key and the public key.

- *Signing*: This algorithm is used to generate the digital signature based on the secret key and the message that needs to be signed.

- *Verifying*: This algorithm is used to verify the authenticity of the message, based on the message itself, the public key, and the signature generated previously. The result of this algorithm is "accept" or "reject."

147

© Stefania Loredana Nita and Marius Iulian Mihailescu 2022
S. L. Nita and M. I. Mihailescu, *Cryptography and Cryptanalysis in Java*,
https://doi.org/10.1007/978-1-4842-8105-5_11

Digital signatures were first mentioned by Diffie and Hellman in [1], but their rigorous description and security requirements were presented by Goldwasser, Micali, and Rivest in [2].

Digital signatures can be stand-alone algorithms or a combination of an encryption algorithm with hash functions. Beware of the second approach as using only certified algorithms for this purpose is recommended. Our first example is using the RSA algorithm (learn about the RSA algorithm in Chapter 10) with the hash function SHA-256 (learn more about the hash function in Chapter 8). Listing 11-1 shows the implementation, and Figure 11-1 shows the output. Java implements a class called Signature from the java.security package that contains a wide range of digital signatures, for example, RSA combined with a hash function, NIST's Digital Signature Algorithm (DSA) [3] combined with a hash function, and the Edwards-Curve signature algorithm [4]. You can learn more about signature schemes implemented in Java in [5].

Listing 11-1. Example of Using Digital Signatures

```
1
2    import java.security.*;
3
4    public class DigitalSignature
5    {
6        //the function will help to generate the digital signature based
             on SHA256 and RSA algorithm, using input and a private key
7        public static byte[] GenerateDS(byte[] dataInput, PrivateKey
             prvKey) throws Exception
8        {
9            Signature signSHA256RSA = Signature.getInstance
                 ("SHA256withRSA");
10            signSHA256RSA.initSign(prvKey);
11            signSHA256RSA.update(dataInput);
12            return signSHA256RSA.sign();
13        }
14
15        //the function will generate an asymmetric key pair based on
             SecureRandom class and using RSA algorithm
16        //128 value
```

```
17      public static KeyPair AKPGenRSA(int size) throws Exception
18      {
19              SecureRandom sr = new SecureRandom();
20              KeyPairGenerator kpg = KeyPairGenerator.getInstance("RSA");
21              kpg.initialize(size, sr);
22              return kpg.generateKeyPair();
23      }
24
25      //function for verifying the digital signature based on a
        public key
26      public static boolean CheckDS(byte[] message, byte[]
        verifyingSignature, PublicKey pubKey) throws Exception
27      {
28              Signature signature = Signature.getInstance
                ("SHA256withRSA");
29              signature.initVerify(pubKey);
30              signature.update(message);
31              return signature.verify(verifyingSignature);
32      }
33
34      //main function
35      public static void main(String args[]) throws Exception
36      {
37              String message = "Welcome To Apress. Enjoy learning
                cryptography.";
38              KeyPair kp512 = AKPGenRSA(512);
39              KeyPair kp1024 = AKPGenRSA(1024);
40              KeyPair kp2048 = AKPGenRSA(2048);
41
42                byte[] generatedSignature512 = GenerateDS(message.
                  getBytes(), kp512.getPrivate());
43                byte[] generatedSignature1024 = GenerateDS(message.
                  getBytes(), kp1024.getPrivate());
44                byte[] generatedSignature2048 = GenerateDS(message.
                  getBytes(), kp2048.getPrivate());
```

```
45
46              System.out.println("The message for which the signature
                will be computed is -> " + message);
47              System.out.println("The length of the message is -> " +
                message.length());
48
49              System.out.println("\n");
50
51              System.out.println("The 512-signature is -> \n " +
                encStringToHex(generatedSignature512));
52
53              System.out.println("Status of 512-verification -> " +
                CheckDS(message.getBytes(), generatedSignature512, kp512.
                getPublic()));
54
55              System.out.println("\n");
56
57              System.out.println("The 1024-signature is -> \n " +
                encStringToHex(generatedSignature1024));
58               System.out.println("Status of 1024-verification -> " +
                 CheckDS(message.getBytes(), generatedSignature1024,
                 kp1024.getPublic()));
59
60              System.out.println("\n");
61
62              System.out.println("The 2048-signature is -> \n " +
                encStringToHex(generatedSignature2048));
63              System.out.println("Status of 2048-verification -> " +
                CheckDS(message.getBytes(), generatedSignature2048, kp2048.
                getPublic()));
64          }
65
66      //encoding string to hex value
67      public static String encStringToHex(byte[] generated_signature)
```

```
68          {
69              StringBuffer hexStringValue = new StringBuffer();
70              for (int count = 0; count < generated_signature.length;
                count++)
71              {
72
73                  hexStringValue.append(encByteValuesToHex(generated_
                    signature [count]));
74              }
75              return hexStringValue.toString();
76          }
77
78          public static String encByteValuesToHex(byte value)
79          {
80              char[] hex_digit_value = new char[2];
81              hex_digit_value[0] = Character.forDigit((value >> 4) &
                0xF, 16);
82              hex_digit_value[1] = Character.forDigit((value & 0xF), 16);
83              return new String(hex_digit_value);
84          }
85  }
```

Figure 11-1. *The result of Listing 11-1*

In lines 9–17, the digital signature is created. To do this, the message (dataInput) and the private key used to sign the message are both required. Here, an instance of the Signature class is created (in the previous example, the signature's type is SHA256withRSA), and then the signing process is initialized with the secret key. Next, the message is signed and then returned. In lines 22–28, the pair of the secret key and the public key for the RSA algorithm is generated. Next, in lines 31–38, the signature is verified, based on the message and the public key, following a process similar to the one for signing. In the main function, there are three examples of signing, namely, using RSA with the key length 512, then 1024, and finally 2048. In lines 84–102 there are helper functions that convert a string to a hex value, and a byte value to a hex value, respectively.

The next signature scheme is ElGamal [6], which is based on the discrete logarithm problem. The protocol is described next.

Key generation.

- In this algorithm, the sender picks a large prime integer value p and a primitive root α.

- The sender picks an integer value a, which is the secret key, and then computes the value $\beta \equiv \alpha a \ (mod \ p)$.

- The public key is $pk = (p, \alpha, \beta)$, and the secret key is $sk = a$.

Signing the message.

- The sender picks randomly an integer value k for which $\gcd(k, p - 1) = 1$ and then computes the value $y_1 \equiv \alpha^k (mod \ p)$ and the value $y_2 \equiv (m - ay_1)k^{-1}(mod \ (p - 1))$, where m is the message that needs to be signed.

Verifying the message.

- To verify the message, the receiver verifies if the result of the following "equality" is true: $y_1^{y_2} \cdot \beta^{y_1} \equiv \alpha^m \ (mod \ p)$.

In Listing 11-2 there is a naïve implementation of the ElGamal signature scheme; Figure 11-2 shows the result.

Listing 11-2. ElGamal Signature Scheme

```
1    import java.math.BigInteger;
2    import java.util.Scanner;
3
4    public class ElGamalSignature {
5        public static void main(String[] args) {
6            Scanner sc = new Scanner(System.in);
7            System.out.println("Value of p");
8            BigInteger p=new BigInteger(sc.next());
9            System.out.println("Value of alpha");
10           BigInteger alpha=new BigInteger(sc.next());
11           System.out.println("Value of m");
12           BigInteger m=new BigInteger(sc.next());
13           System.out.println("Value of k");
14           BigInteger k=new BigInteger(sc.next());
15           sc.close();
16
17           signingAlgorithm sign = new signingAlgorithm(p,
                 alpha, k, m);
18
19           System.out.println("Public key: " + p.intValue() + " " +
                 alpha.intValue() + " " + sign.beta.intValue());
20           System.out.println("y1: " + sign.y1 + "; y2: " + sign.y2);
21
22           verifyingAlgorithm verify = new verifyingAlgorithm(m, sign.
                 y1, sign.y2, sign.alpha, sign.beta, p);
23           verify.verify();
24       }
25
26   }
27
28
29   class signingAlgorithm {
30       public BigInteger p, alpha, beta, k, m, y1, y2;
31       private BigInteger a = new BigInteger("3");
```

```
32
33          public signingAlgorithm(BigInteger p, BigInteger alpha,
            BigInteger k, BigInteger m)
34          {
35              this.p = p;
36              this.alpha = alpha;
37              this.k = k;
38              this.m = m;
39
40              computeBeta();
41              computeY1();
42              computeY2();
43          }
44
45          public void computeBeta() {
46              this.beta = alpha.modPow(a, p);
47          }
48
49          public void computeY1() {
50              this.y1 = alpha.modPow(k, p);
51          }
52
53          public void computeY2() {
54              BigInteger invK = k.modInverse(p.subtract(BigInteger.ONE));
55              this.y2 = ((m.subtract(a.multiply(y1))).modInverse
                (p.subtract(BigInteger.ONE)).m ultiply(invK)).mod
                (p.subtract(BigInteger.ONE));
56          }
57      }
58
59
60  class verifyingAlgorithm {
61
62          public BigInteger m, y1, y2, alpha, beta, p;
63
```

```
64        public verifyingAlgorithm(BigInteger m, BigInteger y1, BigInteger
          y2, BigInteger alpha, BigInteger beta, BigInteger p) {
65            this.m = m;
66            this.y1 = y1;
67            this.y2 = y2;
68            this.alpha = alpha;
69            this.beta = beta;
70            this.p = p;
71        }
72
73        public void verify() {
74            BigInteger left = (y1.pow(y2.intValue())).multiply(beta.pow
              (y1.intValue())).mod(p);
75            BigInteger right = alpha.modPow(m, p);
76
77            if (left.compareTo(right) == 0)
78                System.out.println("Signature verified");
79
80              else
81                    System.out.println("Signature missmatch");
82        }
83    }
```

Figure 11-2. *The result of signing a message with ElGamal*

Note that the previous implementation just follows the steps described for the ElGamal signature scheme. However, in practice, it is recommended to use algorithms already implemented in libraries, as they are tested and validated widely. In advanced implementations, the message is hashed using a hash function, and it's working with the hash value of the message instead of the plain message. You can learn more about signature schemes in [7] and [8].

Conclusion

This chapter presented digital signatures, which are important algorithms used to validate the bond between the sender and the message that it signs. The chapter included two implementations of digital signatures: the first is a combination between an encryption scheme (RSA) and a hash function (SHA256), and the second one is the implementation of the ElGamal signature scheme.

References

[1].   Diffie, W., & Hellman, M. (1976). New directions in cryptography. IEEE transactions on Information Theory, 22(6), 644–654.

[2].   Goldwasser, S., Micali, S., & Rivest, R. L. (1988). A digital signature scheme secure against adaptive chosen-message attacks. SIAM Journal on computing, 17(2), 281–308.

[3].   FIPS PUB 186-4. Digital Signature Standard. Available online: `https://nvlpubs.nist.gov/nistpubs/FIPS/NIST.FIPS.186-4.pdf`

[4].   Bernstein, D. J., Duif, N., Lange, T., Schwabe, P., & Yang, B. Y. (2012). High-speed high-security signatures. Journal of cryptographic engineering, 2(2), 77–89.

[5].   Signature Algorithms, `https://docs.oracle.com/en/java/javase/17/docs/specs/security/standard-names.html#signature-algorithms`

[6].   ElGamal, T. (1985). A public key cryptosystem and a signature scheme based on discrete logarithms. IEEE transactions on information theory, 31(4), 469–472.

[7].   Menezes, A. J., Van Oorschot, P. C., & Vanstone, S. A. (2018). Handbook of applied cryptography. CRC press.

[8].   Stinson, D. R. (2005). Cryptography: theory and practice. Chapman and Hall/CRC.

CHAPTER 12

Identification Schemes

Identification schemes are one of the most important types of scheme used in cryptography. There are two entities, namely, the prover and the verifier, in a scheme. The prover tries to prove its identity to the verifier in such a way that the prover cannot be impersonated by an adversary either in the process of demonstrating its identity or after the demonstration. Once the identity is confirmed, not even the verifier can alter the prover's identity. The process of proving the identity may end with acceptance or rejection (termination without acceptance).

A rudimentary technique of identification is using a password in a particular process. This method provides weak authentication. Identification schemes have many applications, such as access control. Access control is an important part of a system because it manages the access to different resources based on the privileges of each type of user within the system. However, the identification schemes are not related only to human users; they can be used for a specific process authentication. For example, a process/application may use a specific port for communication transmission, or some automated task within an application is implemented to do specific operations according to its privileges. Another handy example is related to our daily lives: we use our biometric features (fingerprints, face) to unlock our smartphones.

An identification protocol takes place in real time, showing that the prover is active and operational during the whole process of identification. To identify a prover, the following techniques can be used:

- *Something known*: In this category, passwords, personal identification numbers (PINs), or a secret key can be included.

- *Something possessed*: This category usually includes physical devices such as magnetic-stripe cards, smart cards, password generator devices for one-time passwords, etc.

© Stefania Loredana Nita and Marius Iulian Mihailescu 2022
S. L. Nita and M. I. Mihailescu, *Cryptography and Cryptanalysis in Java*,
https://doi.org/10.1007/978-1-4842-8105-5_12

- *Something inherited*: This category is related to human physical characteristics, namely, biometric features, such as fingerprint, handwriting or handwritten signatures, voice, etc.

Weak authentication is made by using passwords, and there are several approaches. In this case, the passwords can be stored in specific files that are protected from reading and writing. However, the passwords should not be stored in plain text. Instead, the password should be hashed using a particular hash function or hashing technique (see Chapter 8 for more about hash functions). Another approach involves defining rules for passwords, for example using a minimum number of characters, uppercase letters, special symbols, etc. These rules have the purpose of protecting against dictionary attacks, which can easily break predictable passwords. For an additional level of security, the passwords can be, for example, XORed with a specific value, and then on the obtained value is applied a hash function. Finally, the hashed value is stored for further authentication.

The identification schemes that are based on passwords are exposed to the following types of attacks: replay of fixed passwords, exhaustive password search, password-guessing and dictionary attacks, etc.

There are two other techniques similar to passwords: PINs and two-step authentication. One-time passwords (OTPs) are an intermediary technique between weak authentication (passwords) and strong authentication (zero-knowledge proofs); they provide a stronger authentication than passwords. As the name suggests, when the user wants to authenticate, a different password is used each time. For the OTP, there are several ways in which they can be used: a list of OTPs that is shared between the authentication server and the user, OTPs that are updated continuously, and OTPs that use hash functions.

Zero-knowledge proofs are the most secure way to prove an identity, because they provide strong authentication. To identify itself, the prover may use zero-knowledge proofs, and usually, the proof consists of several pairs of challenge-response messages exchanged between parties (prover and verifier). To prove the identity, the verifier challenges the prover, and the prover must respond with values known only by itself that prove its identity. It works like a secret key-public key pair: the prover generates its response based on its secret key, and the verifier checks the response using the prover's public key. When the messages are exchanged, the verifier learns nothing additional about the verifier (has "zero knowledge"), and no additional information about the prover is leaked such that the verifier convinces other entities about the prover's identity. Figure 12-1 shows a general form of message exchange.

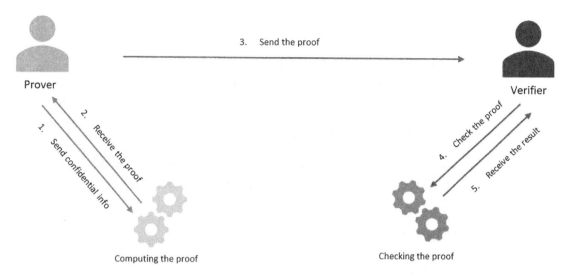

Figure 12-1. *Exchanging the messages between the prover and verifier*

The identification schemes that are based on zero-knowledge proofs have the following characteristics:

- *Completeness*: The message response of the prover should be marked as valid when the claim "prover tries to convince the verifier" is true.

- *Soundness*: The message response of the prover should be marked as valid with a very small probability when the claim "prover tries to convince the verifier" is false.

- *Zero-knowledge*: The verifier does not learn anything additional about the prover in the process of identification.

FFS Identification protocol

An example of a zero-knowledge identification scheme is the Feige-Fiat-Shamir (FFS) identification protocol [4], in which the prover tries to prove its identity to the verifier in several t executions. The protocol has four stages: establishing the system parameters, generating keys for provers, exchanging messages, and taking actions. The protocol FFS is as follows [4]:

- *System parameters*: The trusted authority (TA) chooses two prime numbers p and q, such that p and q are congruent to 3 modulo 4. The two values are kept secret, and the value $n = pq$ is computed and made publicly available. Note that the problem of integer factorization for n should be hard. Considering the probability of impersonation based on the best attack as being 2^{-kt}, then the security parameters for the system are k and t, where t also gives the number of executions.

- *Keys generation*: Every prover should follow these steps:

    1. From the interval $[1, n-1]$, choose randomly k integer values, and call them $s_1, ..., s_k$. Then choose randomly k bit values $b_1, ...b_k$.

    2. Calculate the following value for each

    $$1 \leq i \leq k: v_i = (-1)^{b_i} \bullet \left(s_i^2 \right)^{-1} mod\ n .$$

    3. The public key for the prover is $PK = (v_1, ..., v_k)$, and the secret key is $SK = (s_1, ..., s_k)$.

- *Messages exchange*: For each round t, three messages are exchanged between the prover and the verifier (see the next step for the formulas for each value here):

    1. Prover to verifier: Send the value $x = \pm r^2\ mod\ n$.

    2. Verifier to prover: Send the tuple $e = (e_1, ..., e_k)$, $e_i \in \{0, 1\}$.

    3. Prover to verifier: Send the value $y = r \bullet \prod_{e_j=1} s_j^{e_j}\ mod\ n$.

- *Actions*: For each round t, for the previous values x, e, y some computations should be made. The identity of the prover is accepted if all t rounds terminate with success; otherwise, the identity is rejected. For x, e, y, the computations are as follows:

    1. The prover generates randomly an integer value r, with $1 \leq r \leq n - 1$. Then, the prover generates randomly a bit value b and calculates $x = (-1)^b \bullet r^2\ mod\ n$. The value x is called the *witness*, and it is sent to the verifier.

2. The verifier generates randomly k bit values in the form of $e = (e_1, ..., e_k)$ and sends them to the prover. The values from the tuple are called the *challenge*.

3. The prover calculates the value $y = r \cdot \prod_{e_j=1} s_j^{e_j} \bmod n$ and sends it to the verifier. This value is called the *response*.

4. The verifier calculates the value $z = y^2 \cdot \prod_{j=1}^{n} v_j^{e_j} \bmod n$. Then it checks whether $z = \pm x$ and $z \neq 0$.

Next, the implementation in Java is presented in Listing 12-1, and the result can be checked in Figure 12-2.

Listing 12-1. The Implementation of the FFS Identity Scheme

```
1   import java.math.BigInteger;
2   import java.util.ArrayList;
3   import java.util.BitSet;
4   import java.util.List;
5   import java.util.Random;
6   import java.util.Scanner;
7
8   public class FFSScheme {
9
10      // declare system paramters: prime numbers p,q, computed n,
        and given k
11      private static BigInteger p;
12      private static BigInteger q;
13      private static BigInteger n;
14          private static int k;
15
16      // used to generate random values
17      private static Random random = new Random();
18
19      // the set of k random integer values (private/secret key)
20      private static List<BigInteger> random_s_set = new
        ArrayList<>();
```

```
21          // the set of k random bits
22          private static BitSet random_b_set = new BitSet(k);
23          // the set of k computed values in the form of tuple v
            (public key)
24          private static List<BigInteger> computed_v_set = new
            ArrayList<>();
25
26          // the witness
27          private static BigInteger x;
28          // the set of the challenge values
29          private static BitSet random_e_set = new BitSet(k);
30          // the response
31          private static BigInteger y;
32
33          // the product used in the computation of y
34          private static BigInteger product_s;
35          // the product used in the computation of z
36          private static BigInteger product_v;
37
38          // the random integer value used in messages exchange
39          private static BigInteger r;
40
41          // the random bit value used in messages exchange
42          private static int b;
43
44          // method that implements system parameters generation
45          private static void GenerateParameters() {
46              // p = generate a big prime integer on 128 bits
47              p = BigInteger.probablePrime(128, random);
48              // q = generate a big prime integer on 128 bits
49              q = BigInteger.probablePrime(128, random);
50              // n = pq
51              n = p.multiply(q);
52
53              // k is given by the user
54              System.out.println("k= ");
```

```
55          Scanner sc = new Scanner(System.in);
56          k = sc.nextInt();
57          sc.close();
58      }
59
60      private static void GenerateKeys() {
61
62          for (int i = 0; i < k; i++) {
63
64              // generate each s_i value
65              // the number of bits used for representation of
                each s_i
66              // should be the same as n
67              BigInteger s_i = new BigInteger(n.bitLength() + 1,
                random).mod(n);
68
69              // check whether gcd(s_i, n) = 1
70              while (s_i.gcd(n).intValue() != 1) {
71                  s_i = new BigInteger(n.bitLength() + 1,
                    random).mod(n);
72              }
73
74              // add the value s_i for which gcd(s_i, n) = 1 to
                the s set
75              random_s_set.add(s_i);
76
77              // generate random bit b_i and add it to the set
                of bits b
78              random_b_set.set(i, random.nextBoolean());
79
80              // compute (-1)^(b_i)
81              BigInteger minus_one_pow = (((new BigInteger("-1")).
                pow(random_b_set.get(i) ? 1 : 0)));
82
```

```
83              // compute v_i = (-1)^(b_i) x ((s_i)^2)^(-1) mod n
84              BigInteger computed_v_i = minus_one_pow.multiply
                (random_s_set.get(i).pow(2)).modInverse(n);
85
86              // add v_i to the set v
87              computed_v_set.add(computed_v_i);
88          }
89
90      }
91
92      private static void ExchangeMessages() {
93
94          // generate random integer r
95          // mod n ensures that r is in [1, n-1]
96          r = new BigInteger(n.bitLength() + 1, random).mod(n);
97
98          // generate random bit b
99          b = random.nextBoolean() ? 1 : 0;
100
101         // compute x = (-1)^b x r^2
102         x = (((new BigInteger("-1")).pow(b)).multiply((r.pow(2)))).
            mod(n);
103
104         // generate random bit e_i and add it to the set of
            challenge bits e
105         for (int i = 0; i < k; i++) {
106             random_e_set.set(i, random.nextBoolean());
107         }
108
109         // compute the product (s_1)^(e_1) x ... x (s_k)^(e_k)
110         product_s = new BigInteger("1");
111         for (int i = 0; i < k; i++) {
112             product_s = product_s.multiply(random_s_set.get(i).
                pow(random_e_set.get(i) ? 1 : 0));
113         }
```

```
114
115            // compute y = [the product from above] x r (mod n)
116            y = product_s.multiply(r.mod(n)).mod(n);
117        }
118
119    private static void CheckProof() {
120
121            // compute the product (v_1)^(e_1) x ... x (v_k)^(e_k)
122            product_v = new BigInteger("1");
123            for (int i = 0; i < k; i++) {
124                product_v = product_v.multiply(computed_v_set.get(i).
                   pow(random_e_set.get(i) ? 1 : 0));
125            }
126
127            //compute z = [the product from above] x y^2 (mod n)
128            BigInteger z = (y.pow(2).mod(n)).multiply
                (product_v).mod(n);
129
130            // BigInteger z = new BigInteger("1234555");
131
132            System.out.print("\nz = " + z.toString());
133
134            System.out.print("\nResponse: ");
135
136            //check whether z = +/- x and z != 0
137            if ((z.equals(x) || z.equals(x.negate().mod(n))) &&
                !z.equals(new BigInteger("0"))) {
138                System.out.print("\nAccept");
139            } else {
140                System.out.print("\nReject");
141            }
142        }
143
144    public static void main(String[] args) {
145
```

```
146              GenerateParameters();

147

148              GenerateKeys();

149

150              ExchangeMessages();

151

152              System.out.println("\n=== System parameters and
                 computed values
153   ===");

154

155              System.out.print("\np = " + p);
156              System.out.print("\nq = " + q);
157              System.out.print("\nn = " + n);

158

159              System.out.print("\nSecret key: ( ");
160              for (int i = 0; i < k; i++)
161                  System.out.print(random_s_set.get(i) + " ");
162              System.out.print(")");

163

164              System.out.print("\nPublic key: ( ");
165              for (int i = 0; i < k; i++)
166                  System.out.print(computed_v_set.get(i) + " ");
167              System.out.print(")");

168

169              System.out.print("\nr = " + r);

170

171              System.out.print("\nb = " + b);

172

173              System.out.print("\nx = " + x);

174

175              System.out.print("\nRandom bits e: ( ");
176              for (int i = 0; i < k; i++)
177                  System.out.print(random_e_set.get(i) ? 1 + " " :
                     0 + " ");
178              System.out.print(")");
```

```
179
180                     System.out.print("\ny = " + y);
181
182                 CheckProof();
183             }
184         }
```

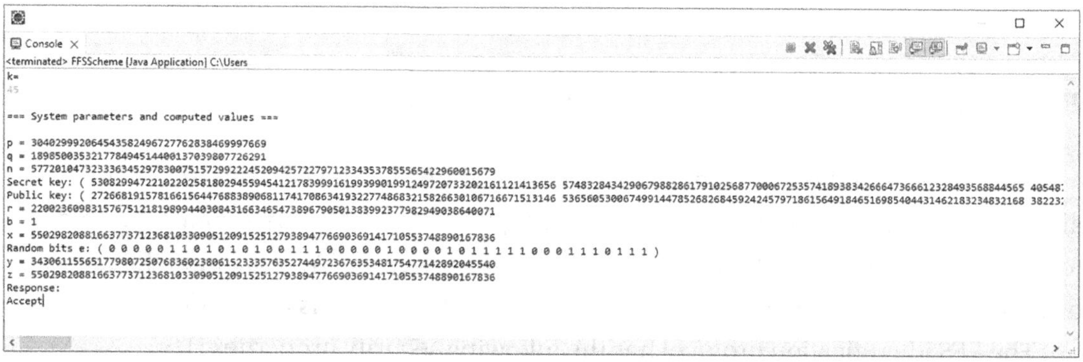

Figure 12-2. *The result for the FFS identification scheme*

The code from Listing 12-1 contains on the first six lines the imported libraries used in the program. The code is divided into four methods, namely, the method that implements the generation of the system parameters, the method for the keys generation, the method that computes the values exchanged between the prover and verifier, and the method that checks whether the response checks the identity of the prover or not. Each method follows the computations presented in the technical details of the FFS scheme. To check if the verification fails when a wrong value is sent, just comment line 132 and uncomment line 134. It is obvious that the integer from line 134 does not fulfill the conditions $z = \pm x$ and $z \neq 0$. Indeed, in this case, the identity is rejected (see Figure 12-3).

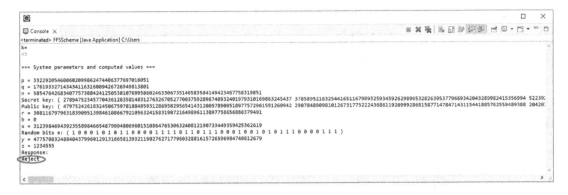

Figure 12-3. *Rejecting the identity in the FFS scheme*

Note that the code from Listing 12-1 is implemented for the $t = 1$ round. For an arbitrary number of rounds, the ExchangeMessages and CheckProof should be called together for that number of times. If each time the message is "Accept," then the identity of the prover is confirmed; otherwise, the identity of the prover is not confirmed.

The FFS identification protocol has the following security properties [1]:

- It is provably secure against the chosen message attacks.

- The hardness assumption lies in the integer factorization problem.

- The verifier learns nothing about the prover in the process of message exchange.

- Well-chosen system parameters give fewer chances of impersonation. For example, choosing the values k, t such that $kt = 20$, the possibility of impersonation is 1 in a million.

- However, if kt is constant while k is increased and t is decreased, it makes the identity scheme be not zero-knowledge anymore.

Conclusion

This chapter showed that there are several types of identification schemes that confirm the identity of an entity, from ones that provide weak authentication (such as passwords) to ones that provide strong authentication (zero-knowledge proofs).

One of the most popular identification protocols is Feige-Fiat-Shamir (FFS), whose hardness assumption lies on the integer factorization problem. This chapter showed how to implement the FFS protocol.

References

[1]. Menezes, A. J., Van Oorschot, P. C., & Vanstone, S. A. (2018). Handbook of applied cryptography. CRC press.

[2]. Stinson, D. R. (2005). Cryptography: theory and practice. Chapman and Hall/CRC.

[3]. Kurosawa, K., & Heng, S. H. (2006, April). The power of identification schemes. In International Workshop on Public Key Cryptography (pp. 364–377). Springer, Berlin, Heidelberg.

[4]. Feige, U., Fiat, A., & Shamir, A. (1988). Zero-knowledge proofs of identity. Journal of cryptology, 1(2), 77–94.

CHAPTER 13

Lattice-Based Cryptography and NTRU

This chapter presents an important topic in cryptography, namely, the encryption schemes based on lattices. These structures are important because it has been proven they can be included in post-quantum cryptography. Lattices are one of the few techniques that can resist quantum attacks and are good candidates for encryption schemes for quantum computers.

Such techniques are important because quantum computers are knocking on the door and the digital world needs to be prepared. Traditional encryption systems, such as RSA, Diffie-Hellmann, and elliptic curves cryptosystems, will be overcome by the technological revolution brought about by quantum computers; therefore, more powerful techniques and tools are necessary to protect the data.

As great as lattices seem, we need to pay a price for using lattices: they are very complex mathematical structures that require solid knowledge of algebra and the ability to understand and work with abstract concepts. Lattices are the basis for a large number of fully homomorphic encryption schemes, which are another hot topic in cryptography nowadays, because they allow computations to be made directly on the encrypted data, obtaining the same result as applying equivalent operations on the plaintext (see Chapter 14 for more details). Examples of lattice-based cryptosystems are Peikert's RLWE Key-Exchange [1], the GGH cryptosystem [2], and the NTRU cryptosystem [3].

This chapter will cover the main mathematical aspects that characterize lattices. The concepts presented represent the minimum theoretical information required to understand lattices.

© Stefania Loredana Nita and Marius Iulian Mihailescu 2022
S. L. Nita and M. I. Mihailescu, *Cryptography and Cryptanalysis in Java*,
https://doi.org/10.1007/978-1-4842-8105-5_13

Consider the n-dimensional real numbers space $\mathbb{R}^n$, whose elements have a form of a row vector $v = (v_1, ..., v_n)$ and each $v_i \in \mathbb{R}$, $i \in \{1, ..., n\}$. A lattice $\mathcal{L}$ has the following representation:

$$\mathcal{L}(v) = \left\{ \sum_{i=1}^{n} a_i v_i \,\middle|\, a_i \in \mathbb{Z} \right\}.$$

Note that the vector v from the previous definition should be a basis in $\mathbb{R}^n$, and a_i is an integer number. In other words, a lattice is the set of all linear combinations with integer coefficients. From the lattice's definition, it can be easily seen that $\mathbb{Z}^n$ is a lattice produced by the standard basis in $\mathbb{R}^n$. Figure 13-1 shows the representation of a lattice in $\mathbb{R}^2$.

Figure 13-1. *Example of a lattice (source:* `https://en.wikipedia.org/wiki/ Lattice_(group)#/media/File:Equilateral_Triangle_Lattice.svg`*)*

Also of interest to cryptography are the computational problems regarding lattices, which lead to the hardness assumptions of cryptosystems. Several computing problems involve lattices: shortest vector problem (SVP), short basis problem (SBP), short independent vector problem (SIVP), or closest vector problem (CVP).

The SVP is as follows: let V be a vector space, b a basis in V, and N a norm. Given a lattice $\mathcal{L}(b)$, a vector $v \in V$ should be found for which its norm in V is the minimum distance in $\mathcal{L}$. Alternatively, v should have the following form:

$$\|v\| = \lambda\big(\mathcal{L}(b)\big),$$

where $\|.\|$ denotes the norm in V and λ is the minimum distance defined in $\mathcal{L}(b)$.

From here, two types of SVP result:

- *Calculational SVP*: Finding the minimum distance $\lambda(\mathcal{L}(b))$, when b and $\mathcal{L}(b)$ are given

- *Decisional SVP*: Deciding whether $\lambda(\mathcal{L}(b)) \leq d$ or $\lambda(\mathcal{L}(b)) > d$, when b, $\mathcal{L}(b)$, and $d \in \mathbb{R}$, $d > 0$ are given

This may seem easy to solve, but when it is working with a large dimension basis, the problem is complicated. For example, in cryptography, the dimensions of the lattices can reach 10,000 instead of just 2 as in the previous example. Therefore, each point will have 10,000 components, and for each component, the conditions from earlier must be fulfilled *simultaneously*.

To test whether an encryption system is quantum-resistant, it must be resistant to attacks based on Shor's algorithm [4].

Among the first cryptosystems that used lattices was the NTRU encryption system [3], which is based on SVP. More exactly, the SVP for NTRU lies in factoring specific polynomials in two polynomials with extremely low coefficients. Those specific polynomials come from a truncated polynomial ring. The algorithms for the NTRU cryptosystem are as follows [3]:

- *Key generation*: Consider $n \in \mathbb{N}$, f, g polynomials with a maximum degree $n - 1$ and with coefficients in the set $\{-1, 0, 1\}$. In addition, for f, two polynomials f_p and f_q should exist such that $f \circ f_p = 1 (mod\ p)$ and $f \circ f_q = 1 (mod\ q)$. The public key is $PK = p \circ f_q \circ g\ (mod\ q)$, and the private key is $SK = (f, f_p, g)$.

- *Encryption*: The message has the form of a polynomial m with coefficients in $\{-1, 0, 1\}$. Generate randomly a polynomial r with low coefficients, keep it private, and encrypt the message as $c = r \circ PK + m\ (mod\ q)$.

- *Decryption*: For decryption, the message is recovered using the computations shown here:

$$a = f \circ c \,(mod\ q)$$

$$b = a \,(mod\ p)$$

$$m = f_p \circ b$$

Practical Implementation of the NTRU Library

The following example of NTRU is based on [5]. The example has been adjusted and commented accordingly to show how the NTRU library can be used in real scenarios by providing encryption and decryption operations over messages. The current library from [5] is used as an extension in the Bouncy Castle library as well. The implementation complies with FIPS regulations for Java 1.7, 1.8, and Java 11 [6].

Listing 13-1 shows the full implementation and commented version of the encryption and decryption operations. In the example provided in Listing 13-1, we have used the net.sf.ntru library, which can be downloaded from `https://github.com/tbuktu/ntru`. To add this as a library to Eclipse, complete the following steps before running the example from Listing 13-1:

1. Right-click the project and navigate to Build Path ➤ Configure
    Build Path.

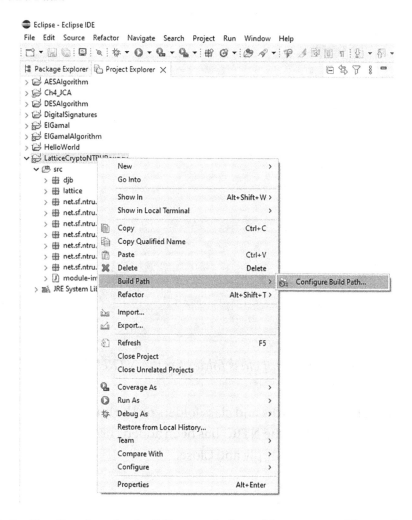

Figure 13-2. *Configuring the building path for adding NTRU library from*
https://github.com/tbuktu/ntru

2. Click Add External Class Folder. Select the path to the location where the NTRU library was downloaded.

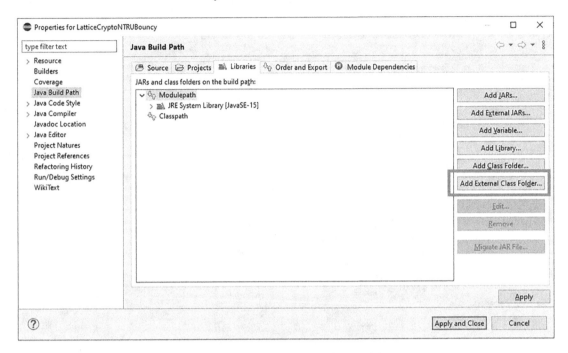

Figure 13-3. *Adding an external class folder for the NTRU library*

3. After adding it to the JARs and class folders on the build path list, check that the path to the NTRU has been added correctly. Once everything is OK, click Apply and Close.

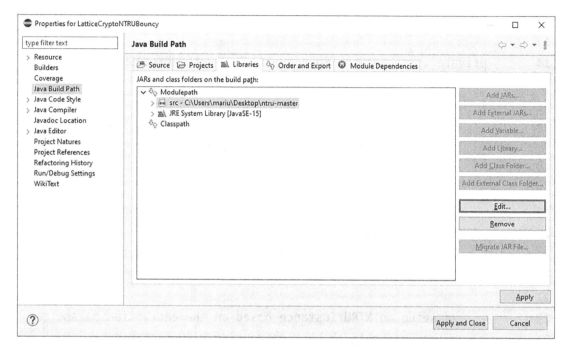

Figure 13-4. *Added NTRU library*

Once steps 1 through 3 have been performed, you should be able to run the example from Listing 13-1 smoothly without any issues.

Listing 13-1. Full Implementation of NTRU Operations, Encryption and Decryption

```
1    import java.util.Arrays;
2    import javax.crypto.*;
3    import javax.crypto.spec.*;
4    import net.sf.ntru.encrypt.*;
5
6    public class AesNtru
7    {
8
9        public static void main(String[] args) throws Exception
10       {
11           new ExampleOf_AES_NTRU().ExecuteNTRUEncryptionDecryption
             Process();
```

```
12        }
13
14        private void ExecuteNTRUEncryptionDecryptionProcess() throws
          Exception
15        {
16            String message = "Welcome to Apress. Enjoy learning " + "
              practical cryptography and NTRU operations";
17
18            // setup the parameters for NTRU and AES
19            String ntru_aes_parameters = "AES/CBC/PKCS5Padding";
20            int length_for_aes = 128;
21            EncryptionParameters parameters_for_ntru = Encryption
              Parameters.APR2011_439_FAST;
22
23            // setup an NTRU instance based on the encryption parameters
24            // and generate the key pair using NTRU instance
25            NtruEncrypt encryption_with_ntru = new NtruEncrypt
              (parameters_for_ntru);
26            EncryptionKeyPair ntru_key_pair = encryption_with_ntru.
              generateKeyPair();
27
28
29            System.out.println("Decrypted message = " + message.
              substring(0, 50) + "...");
30            System.out.println("Length of plain message = " + message.
              length());
31            System.out.println("Maximum length of NTRU = " + parameters_
              for_ntru.getMaxMessageLength());
32
33            // compute the encryption of the message
34            byte[] encryption_message = EncryptTheMessage(message.
              getBytes(),
35                        ntru_key_pair.getPublic(),
36                        ntru_aes_parameters,
37                        length_for_aes,
```

```
38                          parameters_for_ntru);
39
40          System.out.println("Encrypted length = " + encryption_
            message.length +
41                  " (NTRU=" + parameters_for_ntru.
                    getOutputLength() + ", "
42                  + "AES=" + (encryption_message.length - parameters_
                    for_ntru.getOutputLength()) + ")");
43
44          // compute the decryption of the message
45          String decryption_message = new String(DecryptTheMessage
            (encryption_message,
46                  ntru_key_pair,
47                  ntru_aes_parameters,
48                  length_for_aes,
49                  parameters_for_ntru));
50
51        System.out.println("The decryption of message is   = " +
          decryption_message.substring(0, 50) + "...");
52        System.out.println("The length of the decrypted message is = " +
          decryption_message.length());
53      }
54
55      // encryption function will receive the following parameters:
56      // - the public key
57      // - the mode of AES
58      // - the length of AES
59      // - the encryption parameters for NTRU block
60      private byte[] EncryptTheMessage(byte[] clearMessage,
61              EncryptionPublicKey public_key,
62              String modeOfAES,
63              int lengthOfAES,
64              EncryptionParameters ntru_parameters) throws Exception
65      {
66          // compute cryptographic AES key
67          SecretKey cryptoKeyForAES = generateAesKey(lengthOfAES);
```

```
68
69          // generate key specifications for encoding with AES - also
            it will for generating
70          // the initialization vector (IV)
71          SecretKeySpec key_specifications_aes = new SecretKeySpec
            (cryptoKeyForAES.getEncoded(), "AES");
72
73          // providing encryption for message using AES
74          Cipher algorithm = Cipher.getInstance(modeOfAES);
75          algorithm.init(Cipher.ENCRYPT_MODE, key_specifications_aes);
76          byte[] initialization_vector = algorithm.getParameters().
            getParameterSpec(IvParameterSpec.class).getIV();
77          byte[] encryption_with_aes = algorithm.doFinal(clearMessage);
78
79          // encrypt AES key and IV with NTRU
80          NtruEncrypt encryption_with_ntru = new NtruEncrypt(ntru_
            parameters);
81          byte[] cryptoAESKey_Array = cryptoKeyForAES.getEncoded();
82          byte[] initializationVector_and_cryptoKey = generate_byte_
            array(cryptoAESKey_Array, initialization_vector);
83          byte[] encryptedResultWithNtru = encryption_with_ntru.
            encrypt(initializationVector_and_cryptoKey, public_key);
84
85          // put everything in one byte array
86          return generate_byte_array(encryptedResultWithNtru,
            encryption_with_aes);
87      }
88
89
90      // decryption function will receive the following parameters
91      // - the encrypted message as an array of byte
92      // - the decryption key pair
93      // - the mode of AES
94      // - the length of AES
95      // - the encryption parameters related to NTRU block
```

```
96    private byte[] DecryptTheMessage(byte[] encrypted_message,
97            EncryptionKeyPair key_pair_for_encryption,
98            String modeOfAES,
99            int lengthOfAES,
100           EncryptionParameters ntru_parameters) throws Exception
101   {
102       // set the encrypted ntru block based on the NTRU parameters
103       NtruEncrypt encrypted_ntru_block = new NtruEncrypt
          (ntru_parameters);
104
105       // obtain crypto key and initialization vector by decrypting
          the NTRU block
106       byte[] encrypted_block_with_ntru = Arrays.copyOf(encrypted_
          message, ntru_parameters.getOutputLength());
107       byte[] arrayOfKeyAndIV = encrypted_ntru_block.decrypt
          (encrypted_block_with_ntru, key_pair_for_encryption);
108       byte[] arrayOfCryptoAESKey = Arrays.copyOf(arrayOfKeyAndIV,
          lengthOfAES/8);
109       byte[] initializationVectorArray = Arrays.copyOfRange
          (arrayOfKeyAndIV, lengthOfAES/8, 2*lengthOfAES/8);
110
111       // based on the AES crypto key and initialization vector,
          perform the decryption of the message
112       byte[] encrypted_message_with_aes = Arrays.copyOfRange
          (encrypted_message,
113               encrypted_block_with_ntru.length,
114               encrypted_message.length);
115
116       // specify the encryption mode of AES
117       Cipher algorithm = Cipher.getInstance(modeOfAES);
118
119       // configure the key specification related to the algorithm
          that we are using (e.g., AES)
120       SecretKeySpec key_specification_aes = new SecretKeySpec(array
          OfCryptoAESKey, "AES");
```

```
121
122         // set the specifications of the parameters for the
            initialization vector
123         IvParameterSpec initialization_vector_specifications = new Iv
            ParameterSpec(initializationVectorArray);
124
125
126         // initialize the algorithm for decryption by specifying the
            mode, the AES key and initialization vector
127         algorithm.init(Cipher.DECRYPT_MODE, key_specification_aes,
            initialization_vector_specifications);
128
129         // obtain the message in clear based on the encrypted message
            with AES
130         byte[] messageInClear = algorithm.doFinal(encrypted_message_
            with_aes);
131
132         // return the clear version of the message
133         return messageInClear;
134     }
135
136     private SecretKey generateAesKey(int number_of_bits) throws
        Exception
137     {
138          // generate the key for AES
139         KeyGenerator generatorForAESKey = KeyGenerator.getInstance
            ("AES");
140
141         // initialize the key based on the number of the bits
142         generatorForAESKey.init(number_of_bits);
143
144         // get and return the generated crypto key
145         return generatorForAESKey.generateKey();
146     }
147
```

```
148    // based on two arrays (byteArray1 and byteArray2) generate a
       third one (byteArray3) by concatenate
149    // the encrypted result with NTRU and encrypted result with AES
150    private byte[] generate_byte_array(byte[] byteArray1, byte[]
       byteArray2)
151    {
152        // the final result of concatenation of byteArray1 (encrypted
           result with NTRU)
153        // and byteArray2 (encrypted result with AES)
154        byte[] byteArray3 = new byte[byteArray1.length + byteArray2.
           length];
155
156        // perform the concatenation
157        System.arraycopy(byteArray1, 0, byteArray3, 0, byteArray1.
           length);
158        System.arraycopy(byteArray2, 0, byteArray3, byteArray1.
           length, byteArray2.length);
159
160        // return the third-byte array containing the result of the
           concatenation
161        return byteArray3;
162        }
163    }
```

When running the example from Listing 13-1, the output should look like Figure 13-5.

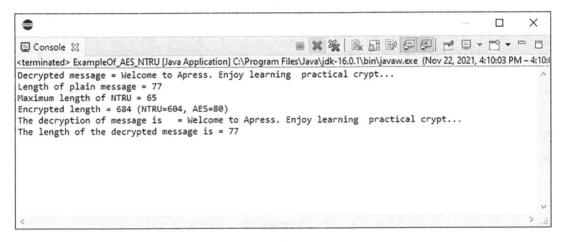

Figure 13-5. *The output of the NTRU encryption and decryption process*

Except the main function, which executes the NTRU encryption and decryption mechanisms, the process is based on these five functions:

```
private void ExecuteNTRUEncryptionDecryptionProcess()
private byte[] EncryptTheMessage()
private byte[] DecryptTheMessage()
private SecretKey generateAesKey()
private byte[] generate_byte_array()
```

In the following paragraphs, we will examine only three functions based on their parameters and purpose. The other two functions are quite self-explanatory. The scenario is pretty simple and straightforward; the main point for the NTRU implementation is represented by the ExecuteNTRUEncryptionDecryptionProcess() function, as shown in Listing 13-2.

The first step is represented in line 3 where we are declaring a variable that will hold the message that we want to encrypt. To achieve the encryption process, we will need to set the parameters for NTRU and AES as mentioned in line 6. In line 7, we will set the length of the encryption key to 128 bits, and in line 8 we will mention through EncryptionParameters the parameter APR2011_439_FAST. Once the parameter is set, it will give 128 bits of security. The mode is designed to work using product-form polynomials [7]. Moving forward, in lines 10 and 11, we set up an NTRU instance using the encryption parameters and cryptographic key generated earlier using the NTRU instance and its given parameters (line 8). Next, in line 14 to 16 we will show different aspects related to the process (as shown in Figure 13-1, the first three lines from the console). In line 18 we declare a byte array for the encrypted message, and the encryption of the message is obtained using the EncryptTheMessage() method (described and shown in Listing 13-3). This process has to be done with respect to the parameters, such as the message (its representation as bytes), the public key from the key pair generated in line 11, the NTRU AES parameters set in line 6, the length of an encryption key for AES (set in line 7), and the parameters mentioned for NTRU (set in line 8). In line 24 we will show in the console the result (as shown in Figure 13-1, the fourth line from the console output). Continue with line 29, we will process the decryption process for the encrypted message with the help of the DecryptTheMessage() method, shown and described in Listing 13-4. The parameters received by the DecryptTheMessage() method are the same as the ones from the EncryptTheMessage method, except the first parameter, which is represented by the encryption form of the message. In lines 35 and 37, we will show the result, as displayed in Figure 13-1, which are the last two lines from the output.

Listing 13-2. ExecuteNTRUEncryptionDecryptionProcess() Function

```
1    private void ExecuteNTRUEncryptionDecryptionProcess() throws Exception
2    {
3            String message = "Welcome to Apress. Enjoy learning " + "
             practical cryptography and NTRU operations";
4
5            String ntru_aes_parameters = "AES/CBC/PKCS5Padding";
6            int length_for_aes = 128;
7            EncryptionParameters parameters_for_ntru = Encryption
             Parameters.APR2011_439_FAST;
8
9            NtruEncrypt encryption_with_ntru = new NtruEncrypt(parameters_
             for_ntru);
10           EncryptionKeyPair ntru_key_pair = encryption_with_ntru.
             generateKeyPair();
11
```

```
12          System.out.println("Decrypted message = " + message.
            substring(0, 50) + "...");
13          System.out.println("Length of plain message = " + message.
            length());
14          System.out.println("Maximum length of NTRU = " + parameters_
            for_ntru.getMaxMessageLength());
15
16          byte[] encryption_message = EncryptTheMessage(message.
            getBytes(),
17                  ntru_key_pair.getPublic(),
18                  ntru_aes_parameters,
19                  length_for_aes,
20                  parameters_for_ntru);
21
22          System.out.println("Encrypted length = " + encryption_message.
            length +
23                  " (NTRU=" + parameters_for_ntru.getOutput
                    Length() + ", "
24                  + "AES=" + (encryption_message.length - parameters_
                    for_ntru.getOutputLength()) + ")");
25
26          String decryption_message = new String(DecryptTheMessage
            (encryption_message,
27                  ntru_key_pair,
28                  ntru_aes_parameters,
29                  length_for_aes,
30                  parameters_for_ntru));
31
32          System.out.println("The decryption of message is   = " +
            decryption_message.substring(0, 50) + "...");
33          System.out.println("The length of the decrypted message is = "
            + decryption_message.length());
34      }
```

We will continue our explanation with the EncryptTheMessage method, which is shown in Listing 13-3 and explained further here. The method is private, and the returning value is represented by a byte array that represents the encrypted message.

In line 1 we have the signature of the method and its parameters/arguments, as follows:

- clearMessage: Represents the message in clear plaintext

- public_key: Represents the cryptographic public key

- modeOfAES: Represents the mode chosen for AES (set in line 6 from Listing 13-2)

- lengthOfAES: Represents the length of the cryptography key for AES in bits (set in line 7 from Listing 13-2)

- ntru_parameters: Represents the NTRU parameters (set in line 8 from Listing 13-2).

Based on these parameters, the processes in Listing 13-3 are executed by following a simple, logical, and rational procedure. In line 7 we will compute the AES cryptography key using the length received as a parameter. In line 9 we use the key specifications to help us to provide the proper encoding using AES, and we will obtain later the initialization vector, which is very important for AES. Further, in line 12 we mention an instance of class Cipher by providing the operating mode for AES. In line 14 we compute and obtain the initialization vector. Lines 18 to 23 represent the encryption process using the AES cryptography key and initialization vector with NTRU.

Listing 13-3. The EncryptTheMessage Method

```
1   private byte[] EncryptTheMessage(byte[] clearMessage,
2               EncryptionPublicKey public_key,
3               String modeOfAES,
4               int lengthOfAES,
5               EncryptionParameters ntru_parameters) throws Exception
6   {
7       SecretKey cryptoKeyForAES = generateAesKey(lengthOfAES);
8
9       SecretKeySpec key_specifications_aes = new SecretKeySpec
        (cryptoKeyForAES.getEncoded(), "AES");
```

```
10
11          Cipher algorithm = Cipher.getInstance(modeOfAES);
12          algorithm.init(Cipher.ENCRYPT_MODE, key_specifications_aes);
13          byte[] initialization_vector = algorithm.getParameters().
            getParameterSpec(IvParameterSpec.class).getIV();
14          byte[] encryption_with_aes = algorithm.doFinal(clearMessage);
15
16           NtruEncrypt encryption_with_ntru = new NtruEncrypt
             (ntru_parameters);
17          byte[] cryptoAESKey_Array = cryptoKeyForAES.getEncoded();
18          byte[] initializationVector_and_cryptoKey = generate_byte_
            array(cryptoAESKey_Array, initialization_vector);
19          byte[] encryptedResultWithNtru = encryption_with_ntru.encrypt
            (initializationVector_and_cryptoKey, public_key);
20
21          return generate_byte_array(encryptedResultWithNtru,
            encryption_with_aes);
22      }
```

Next, we will take a closer look at the DecryptTheMessage method, which is shown in Listing 13-4. The parameters received in line 1 through the signature of the method are similar to the ones from the EncryptTheMessage method shown in Listing 13-3.

Based on the NTRU parameters discussed, in line 7 we need those parameters to provide the proper decryption process. Lines 9 to 15 represent the obtaining process of the cryptographic key and initialization vector using the NTRU block. Moving forward, in line 18, based on the AES cryptographic key and initialization vector, we will perform the decryption of the message. In line 21 we will get an instance to set up the algorithm used in the decryption process and initialized in the variable modeOfAES. Further, in line 23, we will need to specify the key specifications that are related to the algorithm that we are using (e.g., AES); in line 25, we set the specification parameters for the initialization vector; and in line 28, we initialize the algorithm for decryption by specifying the mode, the cryptographic key, and the initialization vector. Moving forward, in line 31 we compute the decryption process and obtain the plaintext, and in line 33 we return the message.

Listing 13-4. DecryptTheMessage Method

```
1    private byte[] DecryptTheMessage(byte[] encrypted_message,
2                EncryptionKeyPair key_pair_for_encryption,
3                String modeOfAES,
4              int lengthOfAES,
5                EncryptionParameters ntru_parameters) throws Exception
6      {
7          NtruEncrypt encrypted_ntru_block = new NtruEncrypt(ntru_
           parameters);
8
9          byte[] encrypted_block_with_ntru = Arrays.copyOf(encrypted_
           message, ntru_parameters.getOutputLength());
10         byte[] arrayOfKeyAndIV = encrypted_ntru_block.
           decrypt(encrypted_block_with_ntru, key_pair_for_encryption);
11         byte[] arrayOfCryptoAESKey = Arrays.copyOf(arrayOfKeyAndIV,
           lengthOfAES/8);
12         byte[] initializationVectorArray = Arrays.copyOfRange
           (arrayOfKeyAndIV, lengthOfAES/8, 2*lengthOfAES/8);
13
14         byte[] encrypted_message_with_aes = Arrays.copyOfRange
           (encrypted_message,
15                  encrypted_block_with_ntru.length,
16                  encrypted_message.length);
17
18         Cipher algorithm = Cipher.getInstance(modeOfAES);
19
20         SecretKeySpec key_specification_aes = new SecretKeySpec
           (arrayOfCryptoAESKey, "AES");
21
22         IvParameterSpec initialization_vector_specifications = new IvP
           arameterSpec(initializationVectorArray);
```

```
23
24          algorithm.init(Cipher.DECRYPT_MODE, key_specification_aes,
            initialization_vector_specifications);
25
26          byte[] messageInClear = algorithm.doFinal(encrypted_message_
            with_aes);
27
28          return messageInClear;
29      }
```

Conclusion

This chapter presented lattices, one of the most important techniques in cryptography. Their importance in quantum cryptography was highlighted, as well as the reasons why lattices are one of the best candidates for post-quantum encryption schemes.

The technical presentation of lattices and the NTRU encryption scheme was followed by a practical example of using NTRU in Java.

References

[1].   Peikert, C. (2014, October). Lattice cryptography for the internet. In International workshop on post-quantum cryptography (pp. 197–219). Springer, Cham.

[2].   Goldreich, O., Goldwasser, S., & Halevi, S. (1997, August). Public-key cryptosystems from lattice reduction problems. In Annual International Cryptology Conference (pp. 112–131). Springer, Berlin, Heidelberg.

[3].   Hoffstein, J., Pipher, J., & Silverman, J. H. (1998, June). NTRU: A ring-based public key cryptosystem. In International Algorithmic Number Theory Symposium (pp. 267–288). Springer, Berlin, Heidelberg.

[4]. Shor, P. W. (1994, November). Algorithms for quantum computation: discrete logarithms and factoring. In Proceedings 35th annual symposium on foundations of computer science (pp. 124–134). IEEE.

[5]. Java Implementation of NTRUEncrypt. GitHub repository: https://github.com/tbuktu/ntru

[6]. Bouncy Castle, FIPS regulations: https://bouncycastle.org/fips_faq.html

[7]. Class SignatureParameters. Available online: http://javadox.com/net.sf.ntru/ntru/1.2/net/sf/ntru/sign/SignatureParameters.html

CHAPTER 14

Advanced Encryption Schemes

Advanced encryption schemes include fully homomorphic encryption (FHE) and searchable encryption (SE), which is a particular case of FHE. More than a decade ago when it was introduced, fully homomorphic encryption was considered a revolutionary approach in cryptography, although privacy homomorphism was introduced in the late 1970s [1]. Both types of encryption schemes are important for today's technologies as they correspond to the current cybersecurity needs. Both are mainly focused on technologies such as cloud computing.

Homomorphic Encryption

Homomorphic encryption is a type of encryption scheme that enables the user to make computations on encrypted data. This eliminates the need for decryption before applying some computations to the data. This saves a lot of time by eliminating additional procedures for downloading the data from the cloud server, decrypting it, applying computations on data, and then encrypting it back and storing it again on the cloud server in encrypted form. A classical use case of homomorphic encryption is when a user who is on a vacation in a foreign country wants to find out which is the nearest restaurant from their geographical position, or where the interesting museums are. A simple search on the Internet for these places can reveal a lot of information about the user: what country they are in, their eating preferences (by looking at the restaurant type), what kind of knowledge they want to learn (by looking at the museum type), their current location, and so on. If the search engine system implemented a homomorphic encryption scheme, then this information about the user would not be accessible to the search engine system anymore, because all search operations would be made over

© Stefania Loredana Nita and Marius Iulian Mihailescu 2022
S. L. Nita and M. I. Mihailescu, *Cryptography and Cryptanalysis in Java*,
https://doi.org/10.1007/978-1-4842-8105-5_14

encrypted data and the user would receive the result in encrypted form. From this simple example, you can see the potential of homomorphic encryption for (almost) all domains of activities.

Let's suppose there are two structures of the same type (groups, rings, or fields): $(G_1, \bot)$, $(G_2, *)$. A function $f: G_1 \rightarrow G_2$ is called homomorphism between G_1, G_2 if

$$f(x_1 \bot x_2) = f(x_1) * f(x_2), \forall x_1, x_2 \in G_1$$

This relation says that the properties of the operation from G_1 are transferred to G_2 via a homomorphism f. In other words, in homomorphic encryption, all computations applied on the plaintext give the same result as applying the corresponding operations (via a homomorphism) over encrypted data. In this case, the result obtained from the plaintext is the same as the decrypted result obtained from the encrypted text. This is translated in technical aspects through the relationship that will be given next. In homomorphic encryption, there is an additional algorithm, besides the well-known key generation, encryption, and decryption. This algorithm is called *evaluation* (usually denoted with *Eval*), and it enables the evaluation of two encrypted texts c_1, c_2 (or an arbitrary number of encrypted text $c_1, ...c_n$) as follows:

$$Dec\left(k_{priv}, Eval_{k_{eval}}(c_1, c_2)\right) = f(m_1, m_2).$$

where m_1, m_2 are the corresponding plain messages from c_1 and c_2. Note that the evaluation algorithm uses an evaluation key, which is public.

From the four mathematical operations, only addition and multiplication should have homomorphic properties, because the *circuit* of any function can be expressed by using only the gates corresponding to these two operations. In short, functions are expressed through logical circuits that contain logical gates (AND, OR, XOR, NOT). More information about logical circuits and how the functions are expressed using them can be found in [2] and [3]. The problem with some homomorphic schemes is that they are constrained by physical limitations, as will be explained in the following categories; therefore, the operations will not be applied unlimitedly.

The intuition for homomorphic encryption is introduced in [1], in the form of privacy homomorphism. However, there are pros and cons to this approach; an example of danger is an attacker that alters the encrypted text, therefore resulting in other plaintext that is not legitimate. The unpadded RSA [4] is considered among the first encryption schemes that support homomorphic properties and works as follows:

$$Enc\left(m_1\right) \bullet Enc\left(m_2\right) = m_1^e m_2^e \ mod \ n = \left(m_1 m_2\right)^e \ mod \ n = Enc\left(m_1 m_2\right),$$

where m_1, m_2 represents plain messages in the form for unpadded RSA and *Enc* is the encryption function of unpadded RSA.

Figure 14-1 shows the history of homomorphic encryption in a graphical form and represents some important encryption systems in each category.

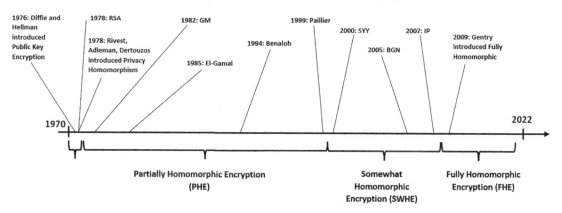

Figure 14-1. *The history of homomorphic encryption*

Homomorphic encryption is divided into three categories, as follows:

- *Partial homomorphic encryption (PHE)*: For this category, one operation of the two (addition, multiplication) is homomorphic, but it can be applied an unlimited number of times. Well-known examples of PHE schemes are RSA [5], Goldwasser-Micali [6], and ElGamal [7]; [5] and [7] are implemented in Chapter 10.

- *Somewhat homomorphic encryption (SWHE)*: This category includes encryption schemes for which both operations are homomorphic, but they can be used a limited number of times. An example of the SWHE scheme is [8].

- *Fully homomorphic encryption (FHE)*: The last and the most exciting category of homomorphic encryption schemes is FHE, which enables both operations to be homomorphic and to be applied an unlimited number of times; therefore, any type of computation can be applied over encrypted data. In the research literature, FHE is considered the "holy grail of cryptography" or the "Swiss Army knife of cryptography" [9]. Gentry's PhD thesis [10] introduced construction for the FHE schemes, using ideal lattices. This was a revolution in cryptography because it opened the door for the FHE. However, Gentry's scheme was very abstract and complex, and it was not practical. The complexity is in the mathematical tools that are used, namely, ideal lattices, and the unpractical aspect is given by the fact that with every evaluation a noise is introduced, and at some time this noise increases so much that the decryption is not possible anymore. FHE itself is divided into three generations: the original scheme, the bootstrapping schemes (that refresh the ciphertext at some point and eliminate the noise), and the schemes based on asymmetric multiplication for which $c_1 \bullet c_2 \neq c_2 \bullet c_1$, although both parts of inequality encrypt the product of the plain corresponding bits $b_1 \bullet b_2$.

Regarding homomorphic encryption, nowadays the focus is on FHE systems, because they can be quantum-resistant. This fact is due to their mathematical foundations, for example, lattices, ring-learning with errors, and NTRU, which are all resistant to Shor's algorithm. However, these mathematical tools are complex and exceed the limitations of physical devices nowadays; therefore, at the moment of writing this book, there is no practical implementation for an FHE scheme that works unlimitedly for any encrypted data. Attempts of FHE implementations exist and lead to powerful FHE libraries, but these are limited and cannot use the entire potential of the FHE schemes that they implement. The following are the FHE libraries, implemented in different programming languages:

- *HElib [11], [12]*: This was developed by IBM and is one of the first implementations of FHE schemes; it implements the BGV scheme [13]. It has a good performance and is one of the few that implements the bootstrapping operation for the BGV scheme. HElib is written in C++.

- *SEAL [14]*: This was developed by Microsoft and implements the BFV [15] and CKKS [16] encryption schemes. It can be used as a library in C# and C++.

- *TFHE [17]*: This implements efficient bootstrap operations and enables functions to be applied directly to the encrypted data.

- *PALLISADE [18]*: This library implements cryptographic primitives for ideal lattices.

For Java, there is a library that implements homomorphic encryption schemes that can be found at [21], but it is deprecated.

HE is standardized through the efforts of actors from different areas of business, such as industry, government, and academic institutions. All resources regarding HE standardization can be found at [19].

Searchable Encryption

Searchable encryption is a particular case of fully homomorphic encryption that allows search operations to be made over encrypted data. A classic example here is the medical unity that needs to retrieve a medical file for a specific patient from the servers by typing the personal Social Security number of the patient. Searchable encryption would work here as the search would be made directly on encrypted data; therefore, there is no need to retrieve the encrypted documents from the server or decrypt them. Searching for text based on a keyword is also an operation, and because the search is on encrypted data, then SE is a particular case of FHE.

The use-case scenario for SE is when there exists an entity that owns a set of documents $D = \{D_1, ..., D_n\}$, which are described by a set of keywords $K = \{kw_1, ..., kw_m\}$. The owner encrypts the documents using an SE scheme and stores them on a cloud server. Then, when the user wants a document (or more) that contains a specific keyword, the user will compute a special value called a *trapdoor* that the search process will be based on. The trapdoor value is then submitted to the cloud server that performs the search operation directly over the encrypted data and then sends the result to the user, who further decrypts the received documents.

Usually, searchable encryption schemes have two components: the *setup* component and the *retrieval* component. In the setup are operations such as data preprocessing, key generation, encryption, and storing data on the cloud servers, while the retrieval component contains operations such as query submission, search, and document retrieval. Searchable encryption has its specific types of users as follows:

- The *data owner* is the entity that owns the data, encrypts it, and stores it in encrypted form on the cloud server.

- The *data user* is the entity that works with the data. The data user can compute trapdoor values based on the query keyword and submit the queries to the cloud server. Also, the data user decrypts the encrypted result received from the server.

- The *cloud server* stores the encrypted data uploaded by the data owner and makes the search operation based on the trapdoor value received from the data user.

As you can see, the roles are very well established and delimited. In addition to the well-known key generation, encryption, and decryption algorithms, searchable encryption has two (or three) more algorithms based on which search is made. These are as follows:

- BuildIndex: This algorithm builds an index structure for the keywords that describe the documents stored on the server in the encrypted format. The algorithm is run by the data owner.

- Trapdoor: This algorithm computes a value called a *trapdoor* based on the query keyword chosen by the data user. Then, this value is submitted to the cloud server.

- Search: This algorithm searches through the encrypted data based on the trapdoor submitted by the data user. The search algorithm is performed by the cloud server.

Regarding their types, SE schemes can be symmetric searchable encryption (SSE) or public-key searchable encryption (PKSE). SSE, like all symmetric schemes, uses just one key for all operations, namely, the private key, while PKSE uses two types of keys, namely, the public key and the private key.

SE schemes should fulfill some specific security requirements, such as controlled searching, encrypted queries, and query isolation [20].

- *Controlled searching* means that the system does not accept search queries from unauthorized users. Moreover, the search operation made by the server should be applied directly to the encrypted data (without decrypting the data in any way).

- *Encrypted queries* means that the keyword for the search query should always be encrypted before being sent to the server.

- *Query isolation* means that the server learns nothing about the keyword, documents, the correspondence between a keyword and a set of documents, etc. In many studies, the server is described as "honest but curious," meaning that the server does its work as it should, but it can try to find relevant information in the process. Query isolation is also required in this setup for the server.

Among the first SE schemes were [22] and [23], and they are important for the categories in which they are included (SSE and PKSE, respectively) because they are considered primary works.

Two libraries that implement searchable encryption schemes are [24] and [25] .

Conclusion

This chapter presented two of the most important topics of cryptography for today's research community. Fully homomorphic encryption is important because the techniques that it uses and its mathematical properties make it quantum resistant. This is a powerful property as quantum computers already exist in dedicated research laboratories. Searchable encryption is also an important topic, being a particular case of fully homomorphic encryption. Its focus is on searching for specific data that fulfills some criteria, an aspect that is very useful in the database field.

References

[1]. Rivest, R. L., Adleman, L., & Dertouzos, M. L. (1978). On data banks and privacy homomorphisms. Foundations of secure computation, 4(11), pp. 169–180.

[2]. Basic Gates and Functions. Available online: `http://www.ee.surrey.ac.uk/Projects/CAL/digital-logic/gatesfunc/index.html`

[3]. CS 370 Computer Architecture Spring 2020. Available online: `https://taoxie.sdsu.edu/cs370/`

[4].   Rivest, R. L., Shamir, A., Adleman, L. (1978). A method for obtaining digital signatures and public-key cryptosystems. Communications of the ACM, 21(2), pp. 120–126.

[5].   Ronald L. Rivest, Adi Shamir, and Leonard Adleman, "A method for obtaining digital signatures and public-key cryptosystems." Communications of the ACM 21.2 (1978): 120–126.

[6].   Shafi Goldwasser and Silvio Micali, "Probabilistic encryption and how to play mental poker keeping secret all partial information." Proceedings of the Fourteenth Annual ACM Symposium on Theory of Computing. 1982.

[7].   Taher ElGamal, "A public key cryptosystem and a signature scheme based on discrete logarithms." IEEE transactions on information theory 31.4 (1985): 469–472.

[8].   Dan Boneh, Eu-Jin Goh, and Kobbi Nissim, "Evaluating 2-DNF formulas on ciphertexts." Theory of Cryptography Conference. Springer, Berlin, Heidelberg, 2005.

[9].   B. Barak and Z. Brakerski. "The Swiss Army Knife of Cryptography," http://windowsontheory.org/2012/05/01/theswiss-army-knife-of-cryptography/, 2012.

[10].   Craig Gentry, "Fully homomorphic encryption using ideal lattices." Proceedings of the forty-first annual ACM symposium on Theory of computing. 2009.

[11].   Halevi, S., Shoup, V. (2013). Design and implementation of a homomorphic encryption library. IBM Research (Manuscript), Vol. 6, pp. 12–15.

[12].   HElib Documentation, https://homenc.github.io/HElib/

[13].   Brakerski, Z., Gentry, C., & Vaikuntanathan, V. (2014). (Leveled) fully homomorphic encryption without bootstrapping. ACM Transactions on Computation Theory (TOCT), 6(3), pp. 1–36.

[14].   Microsoft SEAL (release 3.2.0). February 2019, http://sealcrypto.org. Microsoft Research, Redmond, WA

[15]. Fan, J., & Vercauteren, F. (2012). Somewhat Practical Fully Homomorphic Encryption. IACR Cryptology ePrint Archive, Report 2012/144.

[16]. Cheon, J. H., Kim, A., Kim, M., & Song, Y. (2017i). Homomorphic encryption for arithmetic of approximate numbers. In International Conference on the Theory and Application of Cryptology and Information Security (pp. 409–437). Springer, Cham.

[17]. Chillotti, I., Gama, N., Georgieva, M., & Izabachene, M. (2016). Faster fully homomorphic encryption: Bootstrapping in less than 0.1 seconds. In international conference on the theory and application of cryptology and information security (pp. 3–33). Springer, Berlin, Heidelberg.

[18]. Polyakov, Y., Rohloff, K., Ryan, G. W. (2017). PALISADE lattice cryptography library user manual. Cybersecurity Research Center, New Jersey Institute of Technology (NJIT), Tech. Rep. `https://git.njit.edu/groups/palisade`

[19]. Albrecht M, Chase M, Chen H, Ding J, Goldwasser S, Gorbunov S, Halevi S, Hoffstein J, Laine K, Lauter K, Lokam S, Micciancio D, Moody D, Morrison T, Sahai A, Vaikuntanathan V. (2018). Homomorphic encryption security standard. Technical report. Toronto, Canada: HomomorphicEncryption.org

[20]. Song, D. X., Wagner, D., Perrig, A. (2000). Practical techniques for searches on encrypted data. In Proceeding 2000 IEEE Symposium on Security and Privacy. SP 2000 (pp. 44–55). IEEE.

[21]. kryptnostic/fhe-core, `https://github.com/kryptnostic/fhe-core`

[22]. E. J. Goh, "Secure indexes." IACR Cryptology ePrint Archive, 2003, 216

[23]. D. Boneh, G. Di Crescenzo, R. Ostrovsky, and G. Persiano, "Public key encryption with keyword search." In International conference on the theory and applications of cryptographic techniques (pp. 506–522). 2004. Springer, Berlin, Heidelberg

[24]. encryptedsystems/Clusion, `https://github.com/encryptedsystems/Clusion`

[25]. Crypteron, `https://www.crypteron.com/`

CHAPTER 15

Cryptography Tools

This chapter presents two of the most popular software applications and tools for cryptography that are great for learning and practice purposes: CrypTool (CT) [1] and OpenSSL [2].

CryptTool

The original version of CT was written in C++, but there are several versions, including an online version and a version written in Java, called JCrypTool (JCT). The latest stable version is 1.0.8, and it is supported on Windows, macOS, and Linux; it can be downloaded from the official website at `https://www.cryptool.org/en/jct/downloads`. After downloading the archive, unarchive it and then launch the executable file called `JCrypTool.exe`. After opening it, you'll see the main window of JCT, as shown in Figure 15-1. Here, there are several options. To start using JCT, click the Start JCT option.

© Stefania Loredana Nita and Marius Iulian Mihailescu 2022
S. L. Nita and M. I. Mihailescu, *Cryptography and Cryptanalysis in Java*,
https://doi.org/10.1007/978-1-4842-8105-5_15

Figure 15-1. *The main window of JCT*

You'll see four main sections after starting to use JCT, as shown in Figure 15-2. The first section is File Explorer, which, by default, shows the content of the Users folder. The second section is the Help window, where you can find descriptions of different functionalities of JCT . The third section is a .txt file that can be written with different input values for a chosen cryptosystem. Lastly, the fourth section contains a list of encryption systems and cryptographic primitives that can be tested; it is called CryptoExplorer. You can choose different operations in CryptoExplorer such as Analysis, Visuals, or Games. For example, if you select an encryption system from the fourth section, then in the Help section you will see information about the selected topic in the window shown in Figure 15-3.

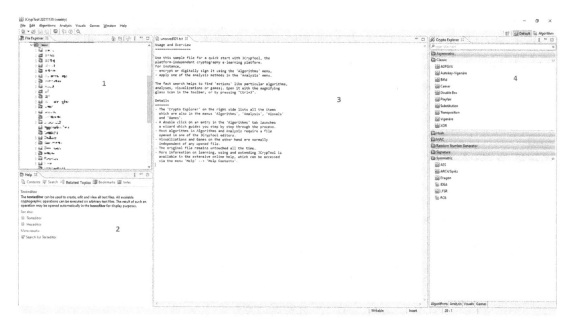

Figure 15-2. *The window sections of JCT*

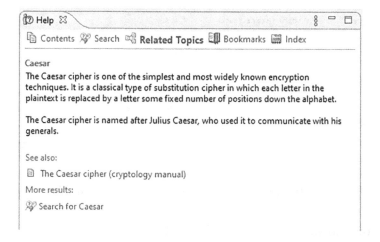

Figure 15-3. *The Help section of JCT*

The first example of using JCT is an easy one; namely, we'll look at the Caesar cipher. To start, edit the text in the third section of the window with a message that will be encrypted using Caesar; then in the Algorithms menu choose Classic ➤ Caesar. Next, a window will open, where you'll choose the secret key and set some settings (Figure 15-4).

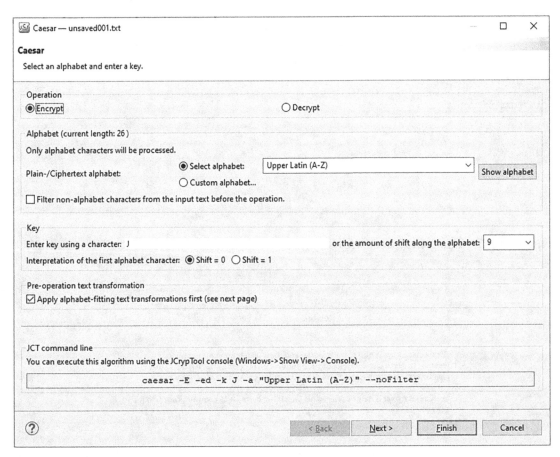

Figure 15-4. *Choosing the secret key and the settings for the Caesar cipher*

In the window shown in Figure 15-4, choose the type of operation, i.e., encryption or decryption. You'll see settings for the alphabet. The default setting is Upper Latin (A-Z), but you can choose from a wide range of alphabets; the tool even supports a user-defined alphabet. The alphabet can be displayed by clicking the "Show alphabet" button, and the input can be filtered, such that the symbols that are not contained in the alphabet are excluded from the input. Then select the key by choosing the index of the symbol key in the alphabet or the letter directly. Here also is the setting for the index of the first character of the alphabet, and it can be set to 0 or 1. The Caesar encryption can be executed from the command line, and in the last setting, the text of the command is shown. Once you've established the encryption key and the settings, then you can click the Next button, which leads to the window in Figure 15-5.

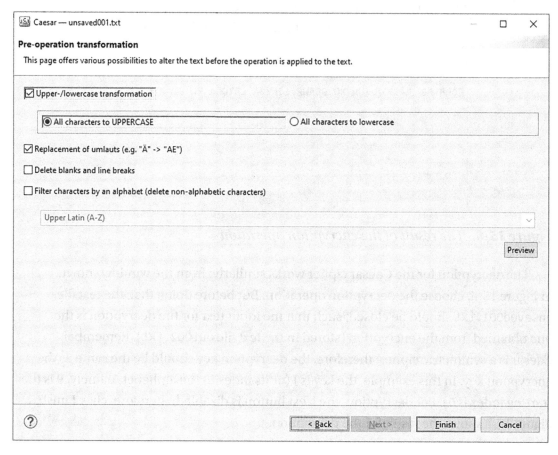

Figure 15-5. Choosing the secret key and the settings for the Caesar cipher

The settings in Figure 15-5 refer to some transformations that can be applied to the input before encryption, for example transforming all characters to uppercase, replacing some special symbols, deleting white spaces, or eliminating the characters that do not belong to the alphabet; then click the Finish button. The result is shown in Figure 15-6, and it is saved in a text file called out002.txt (or a similar name). Note that the encrypted text is displayed in a separate text window. For this example, the text for encryption is "This is an example of using Caesar cipher in JCrypTool."

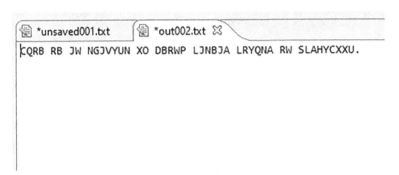

Figure 15-6.   *The result of the encryption operation*

The decryption for the Caesar cipher works similarly: from the window shown in Figure 15-4, choose the decryption operation. But before doing that, the text file unsaved001.txt should be closed, such that the input text for the decryption is the one obtained from the encryption (stored in the text file out002.txt). Remember, Caesar is a symmetric cipher; therefore, the decryption key should be the same as the encryption key. In this example, the key is J (or its index in the alphabet; namely, 9 is the starting index is 0). For decryption, the Next button is disabled; therefore, click Finish. Figure 15-7 shows the result for the decryption.

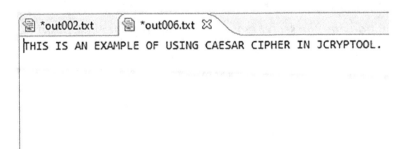

Figure 15-7.   *Decryption for the Caesar cipher*

The next example, of the RSA encryption system, is more complex. Before selecting the type of cryptosystem, type the message to be encrypted into a text file. For this example, the message is "Using RSA to encrypt the message." Choose Algorithms ➤ Asymmetric ➤ RSA from the menu. The window shown in Figure 15-8 opens.

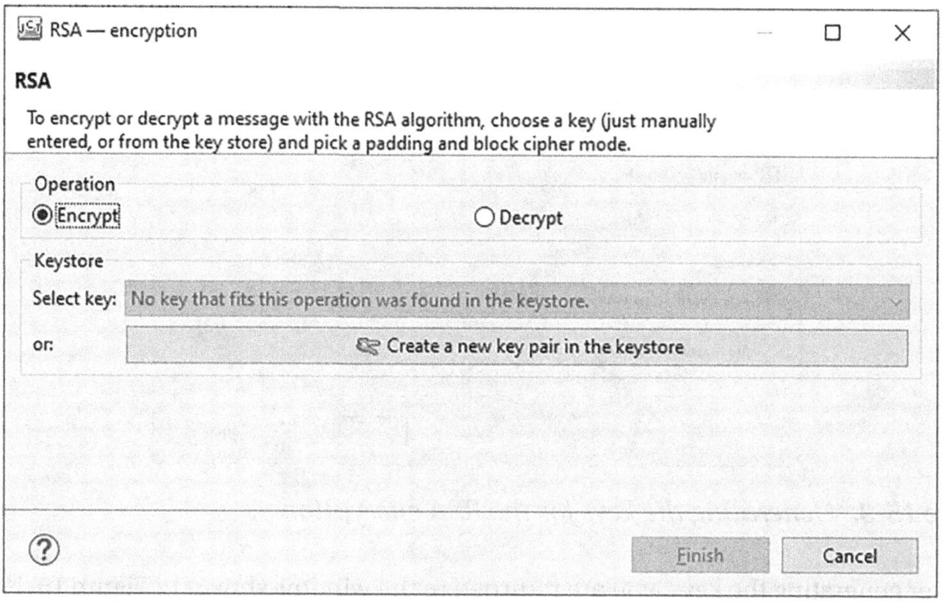

Figure 15-8. *The window for RSA encryption*

From here, choose Encrypt as the operation and then click "Create a new key pair in the keystore," if you do not already have a secret-public key pair for the RSA.

For the new key pair, some information should be completed: the contact name, the algorithm for generating the pair and the length, and, lastly, the passwords for the process. The default contact name is [PW: 1234] Alice Whitehead, but for this example, we have changed it to Apress Reader. Keep the default values for the algorithm and key length. The password for this example is RSA@2021. Then click Finish (see Figure 15-9).

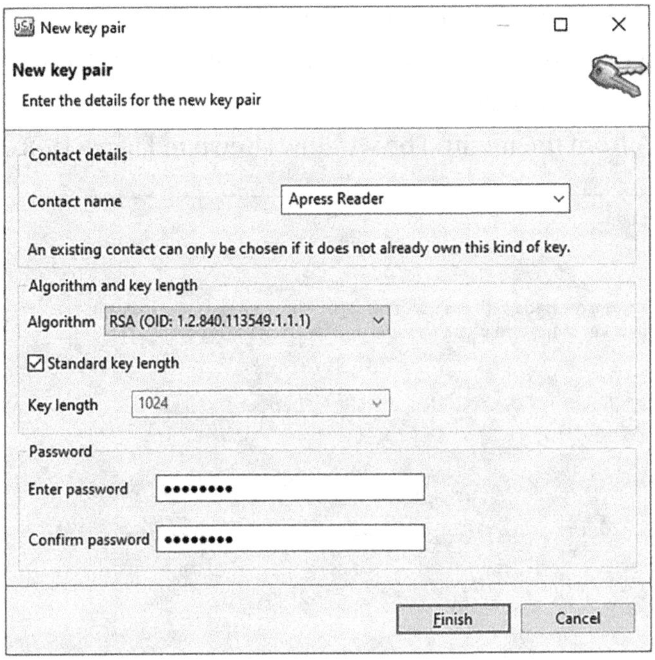

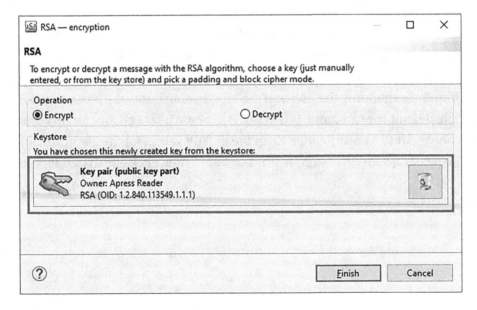

Figure 15-9. *Generating the keys for the RSA encryption*

After generating the keys, you are returned to the window shown in Figure 15-10, where you should click the Finish button.

Figure 15-10. *RSA window after key pair generation*

The output for the encryption process is the encrypted text, which is shown in a new generated file with the extension .bin. The encrypted message from this example looks like the one from Figure 15-11. On the left side of the .bin file is the representation of the message in hexadecimal format, while in the right part there is the message in text form. For a selected byte, some information is shown at the bottom of the window: the hexadecimal value, the decimal, and the binary form. At the bottom left, you'll see the number of the current byte and the total number of bytes in the .bin file.

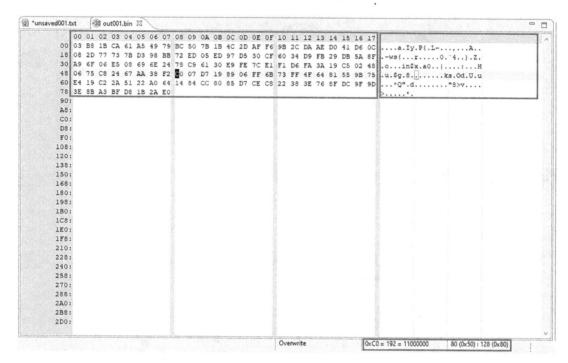

Figure 15-11. *The encrypted message resulted from using the RSA cryptosystem*

To decrypt the message, choose again Algorithms ➤ Asymmetric ➤ RSA. Then, from the open window, choose "Decryption as operation" this time. Note this time the pair of keys previously generated is not preloaded, and the button for generating a new pair is disabled. From here, choose the pair of keys generated previously and then click Finish (Figure 15-12). Next, the password used to create the pair of keys should be introduced; click OK (Figure 15-13). Note it is important to not forget the password used in the key generation.

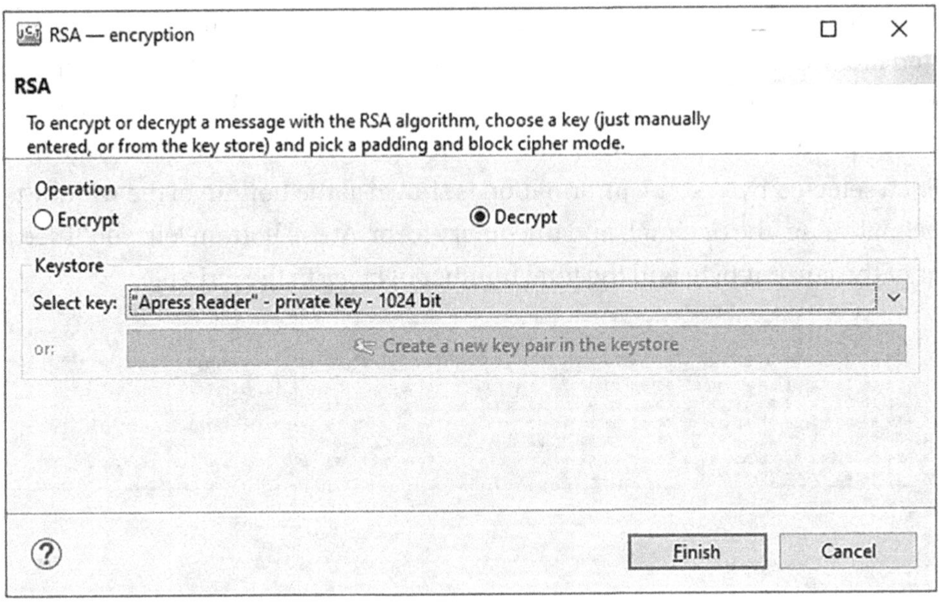

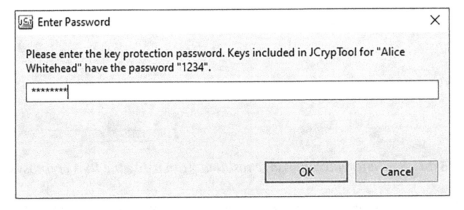

Figure 15-12. *Decryption for the RSA cryptosystem*

Figure 15-13. *Introducing the password for the decryption process*

A similar `.bin` file is generated, which contains the decrypted message in hexadecimal format and text format (Figure 15-14).

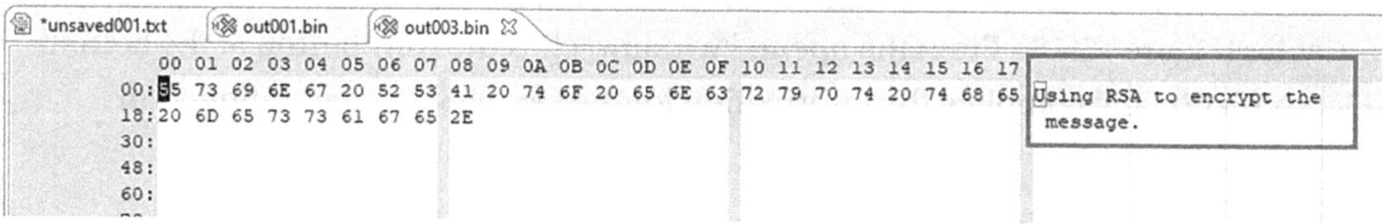

Figure 15-14. *The decrypted message*

CryptTool is a powerful software application for cryptography that can be used for different purposes such as encryption/decryption with a wide range of cryptosystems (symmetric and asymmetric), key generation, hashing, pseudorandom number generation, cryptanalysis, etc.

OpenSSL

OpenSSL is another powerful tool for cryptography. It can be used in different scenarios such as generating keys for different cryptosystems (although the keys can be generated directly with Java) and working with SSL requests. OpenSSL can be downloaded from the official website (`https://slproweb.com/products/Win32OpenSSL.html`). The latest version for Windows at this moment is 3.0.0.

Download the installer, open it, and then follow the installation steps, leaving the default settings (Figure 15-15).

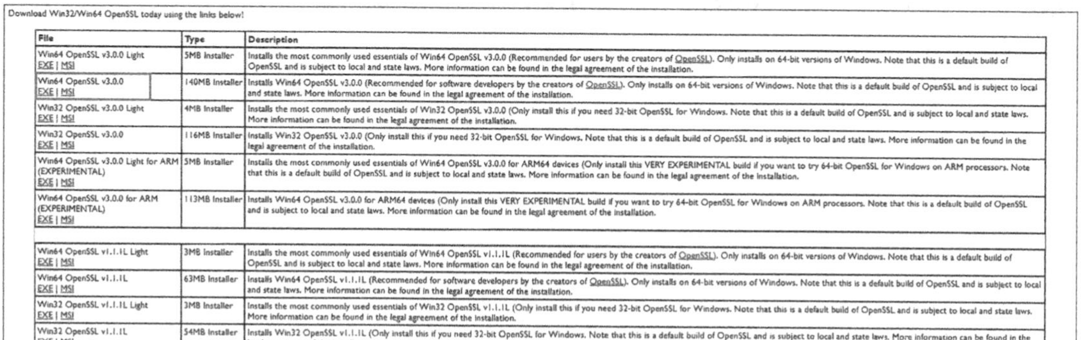

Download Win32/Win64 OpenSSL today using the links below!

File	Type	Description
Win64 OpenSSL v3.0.0 Light EXE | MSI	5MB Installer	Installs the most commonly used essentials of Win64 OpenSSL v3.0.0 (Recommended for users by the creators of OpenSSL). Only installs on 64-bit versions of Windows. Note that this is a default build of OpenSSL and is subject to local and state laws. More information can be found in the legal agreement of the installation.
Win64 OpenSSL v3.0.0 EXE | MSI	140MB Installer	Installs Win64 OpenSSL v3.0.0 (Recommended for software developers by the creators of OpenSSL). Only installs on 64-bit versions of Windows. Note that this is a default build of OpenSSL and is subject to local and state laws. More information can be found in the legal agreement of the installation.
Win32 OpenSSL v3.0.0 Light EXE | MSI	4MB Installer	Installs the most commonly used essentials of Win32 OpenSSL v3.0.0 (Only install this if you need 32-bit OpenSSL for Windows. Note that this is a default build of OpenSSL and is subject to local and state laws. More information can be found in the legal agreement of the installation.
Win32 OpenSSL v3.0.0 EXE | MSI	116MB Installer	Installs Win32 OpenSSL v3.0.0 (Only install this if you need 32-bit OpenSSL for Windows. Note that this is a default build of OpenSSL and is subject to local and state laws. More information can be found in the legal agreement of the installation.
Win64 OpenSSL v3.0.0 Light for ARM (EXPERIMENTAL) EXE | MSI	5MB Installer	Installs the most commonly used essentials of Win64 OpenSSL v3.0.0 for ARM64 devices (Only install this VERY EXPERIMENTAL build if you want to try 64-bit OpenSSL for Windows on ARM processors. Note that this is a default build of OpenSSL and is subject to local and state laws. More information can be found in the legal agreement of the installation.
Win64 OpenSSL v3.0.0 for ARM (EXPERIMENTAL) EXE | MSI	113MB Installer	Installs Win64 OpenSSL v3.0.0 for ARM64 devices (Only install this VERY EXPERIMENTAL build if you want to try 64-bit OpenSSL for Windows on ARM processors. Note that this is a default build of OpenSSL and is subject to local and state laws. More information can be found in the legal agreement of the installation.
Win64 OpenSSL v1.1.1L Light EXE | MSI	3MB Installer	Installs the most commonly used essentials of Win64 OpenSSL v1.1.1L (Recommended for users by the creators of OpenSSL). Only installs on 64-bit versions of Windows. Note that this is a default build of OpenSSL and is subject to local and state laws. More information can be found in the legal agreement of the installation.
Win64 OpenSSL v1.1.1L EXE | MSI	63MB Installer	Installs Win64 OpenSSL v1.1.1L (Recommended for software developers by the creators of OpenSSL). Only installs on 64-bit versions of Windows. Note that this is a default build of OpenSSL and is subject to local and state laws. More information can be found in the legal agreement of the installation.
Win32 OpenSSL v1.1.1L Light EXE | MSI	3MB Installer	Installs the most commonly used essentials of Win32 OpenSSL v1.1.1L (Only install this if you need 32-bit OpenSSL for Windows. Note that this is a default build of OpenSSL and is subject to local and state laws. More information can be found in the legal agreement of the installation.
Win32 OpenSSL v1.1.1L EXE | MSI	54MB Installer	Installs Win32 OpenSSL v1.1.1L (Only install this if you need 32-bit OpenSSL for Windows. Note that this is a default build of OpenSSL and is subject to local and state laws. More information can be found in the legal agreement of the installation.

Figure 15-15. *Official download for OpenSSL*

The next step is to add the OpenSSL configuration and binaries to the system variables.

To run Environment Variables, go to System Properties and click Environment Variables (Figure 15-16). From the opened window, click the New button at the bottom (Figure 15-17), and complete the information for the OpenSSL configuration as in Figure 15-18.

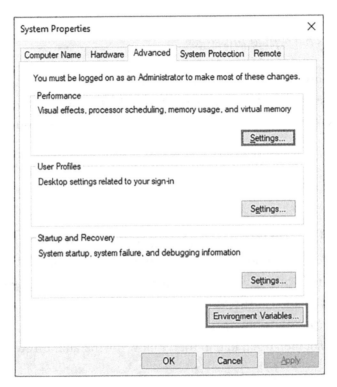

Figure 15-16. *Opening the system variables for editing*

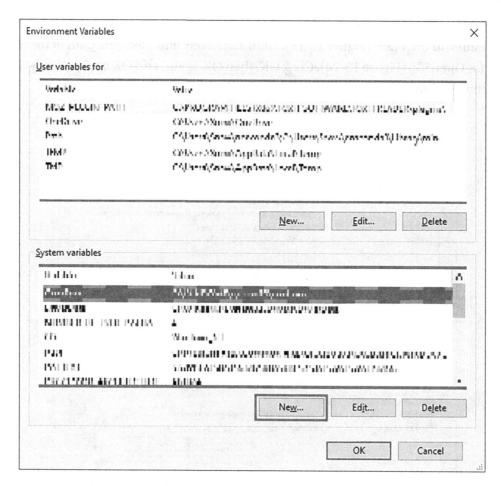

Figure 15-17. *Adding a new entry*

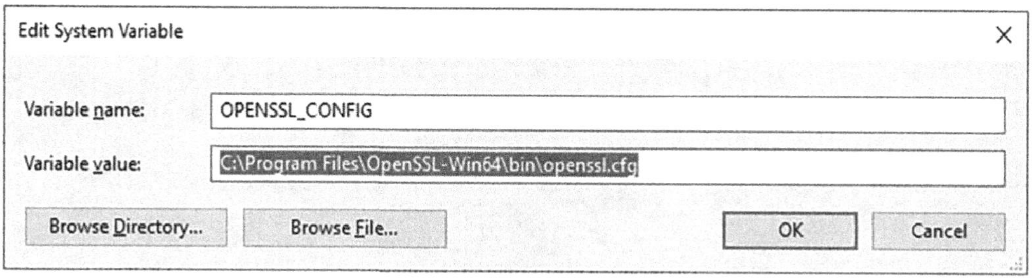

Figure 15-18. *The configuration values for the new entry*

217

Then, add the binary files for OpenSSL, as follows: in the "System variables" section select Path and then Edit (Figure 15-19). Here, click New and paste the path of the bin folder for OpenSSL (Figure 15-20). Click OK, then OK again, and OK again. The System Properties window closes.

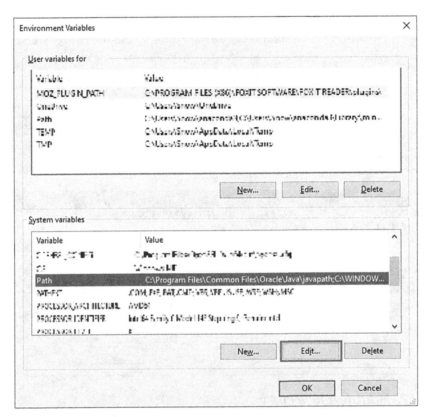

Figure 15-19. *Editing the path from System Variables*

Figure 15-20. *Adding the bin folder for OpenSSL to Path*

To test whether OpenSSL was configured and added properly, open a terminal window and type the command `openssl version`. The version of the library should be displayed (Figure 15-21). If you need help with the OpenSSL commands, just type the `openssl` command in the terminal.

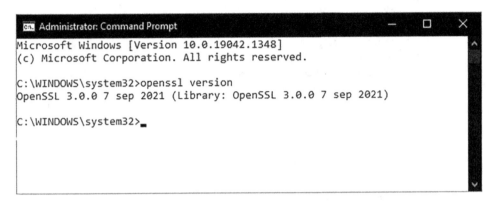

Figure 15-21.   *Testing OpenSSL*

To demonstrate the use of OpenSSL, the next example shows how to generate the private and public keys for the RSA cryptosystem.

First, change the current directory to a specific folder that can be easily found. In this example, we are placing the folder in `C:\openssl`. The command to generate Alice's private key is as follows:

```
openssl genrsa -out alicePrivKey.pem 2048
```

To generate the public key, the command is as follows:

```
openssl rsa -pubout -in alicePrivKey.pem -out alicePublicKey.pem
```

Check the content of the current directory; indeed, the two keys have been generated (Figure 15-22).

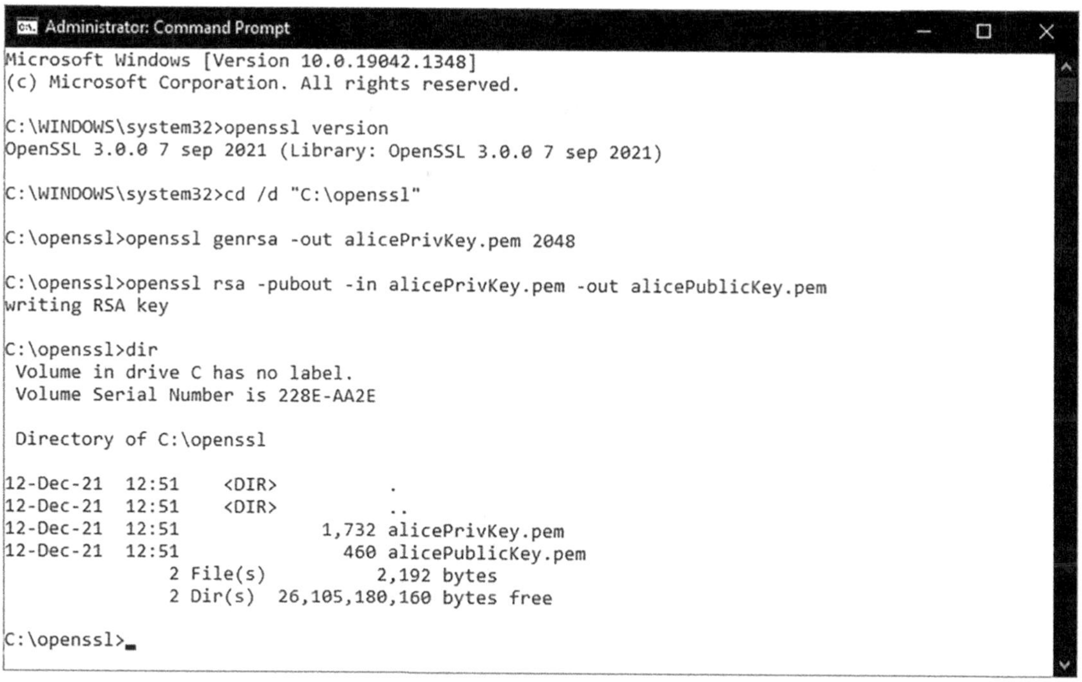

```
Administrator: Command Prompt                                    —   □   ✕

Microsoft Windows [Version 10.0.19042.1348]
(c) Microsoft Corporation. All rights reserved.

C:\WINDOWS\system32>openssl version
OpenSSL 3.0.0 7 sep 2021 (Library: OpenSSL 3.0.0 7 sep 2021)

C:\WINDOWS\system32>cd /d "C:\openssl"

C:\openssl>openssl genrsa -out alicePrivKey.pem 2048

C:\openssl>openssl rsa -pubout -in alicePrivKey.pem -out alicePublicKey.pem
writing RSA key

C:\openssl>dir
 Volume in drive C has no label.
 Volume Serial Number is 228E-AA2E

 Directory of C:\openssl

12-Dec-21  12:51    <DIR>          .
12-Dec-21  12:51    <DIR>          ..
12-Dec-21  12:51             1,732 alicePrivKey.pem
12-Dec-21  12:51               460 alicePublicKey.pem
               2 File(s)          2,192 bytes
               2 Dir(s)   26,105,180,160 bytes free

C:\openssl>_
```

Figure 15-22. *Generating the keys for the RSA cryptosystem*

Conclusion

Cryptography and information security are some of the most exciting and important branches of computer science and information technology as technology evolves and threats may appear anytime even from unexpected sources.

This chapter presented two important software applications and tools used in these fields, namely, CrypTool and OpenSSL. They are very powerful and provide complex operations. Also, the chapter presented how the tools can be installed and how they can be used. The JCrypTool version is even more relevant as it is written in Java, while OpenSSL can be used to parametrize things in Java cryptosystems or for other cryptographic purposes.

References

[1]. CrypTool. Available online: `https://www.cryptool.org/en/`

[2]. OpenSSL. Available online: `https://slproweb.com/`

Index

© Stefania Loredana Nita and Marius Iulian Mihailescu 2022
S. L. Nita and M. I. Mihailescu, *Cryptography and Cryptanalysis in Java*,
https://doi.org/10.1007/978-1-4842-8105-5

Printed in the United States
by Baker & Taylor Publisher Services